6th EEIGM International Conference on Advanced Materials Research

(EEIGM 2011)

IOP Conference Series: Materials Science and Engineering
Volume 31

Nancy, France
7 - 8 November 2011

ISBN: 978-1-63439-842-8
ISSN: 1757-8981

Printed from e-media with permission by:

Curran Associates, Inc.
57 Morehouse Lane
Red Hook, NY 12571

Some format issues inherent in the e-media version may also appear in this print version.

Printed by Curran Associates, Inc. (2015)

For permission requests, please contact the Institute of Physics
at the address below.

Institute of Physics
Dirac House, Temple Back
Bristol BS1 6BE UK

Phone: 44 1 17 929 7481
Fax: 44 1 17 920 0979

techtracking@iop.org

Additional copies of this publication are available from:

Curran Associates, Inc.
57 Morehouse Lane
Red Hook, NY 12571 USA
Phone: 845-758-0400
Fax: 845-758-2634
Email: curran@proceedings.com
Web: www.proceedings.com

Table of contents

Volume 31

6th EEIGM International Conference on Advanced Materials Research
7–8 November 2011, European School of Materials Engineering (EEIGM), Nancy, France

Accepted papers received: 6 February 2012
Published online: 20 February 2012

Papers

Organic and composite materials

Characterization methods

IOP Conf. Series: Materials Science and Engineering **31** (2012) 012001 doi:10.1088/1757-899X/31/1/012001

3-D FEM Modeling of fiber/matrix interface debonding in UD composites including surface effects

A Pupurs and J Varna[1]
Department of Engineering Sciences and Mathematics, Division of Materials Science, Luleå University of Technology, SE 97187, Luleå, Sweden

E-mail: andrejs.pupurs@ltu.se, Janis.Varna@ltu.se

Abstract. Fiber/matrix interface debond growth is one of the main mechanisms of damage evolution in unidirectional (UD) polymer composites. Because for polymer composites the fiber strain to failure is smaller than for the matrix multiple fiber breaks occur at random positions when high mechanical stress is applied to the composite. The energy released due to each fiber break is usually larger than necessary for the creation of a fiber break therefore a partial debonding of fiber/matrix interface is typically observed. Thus the stiffness reduction of UD composite is contributed both from the fiber breaks and from the interface debonds. The aim of this paper is to analyze the debond growth in carbon fiber/epoxy and glass fiber/epoxy UD composites using fracture mechanics principles by calculation of energy release rate G_{II}. A 3-D FEM model is developed for calculation of energy release rate for fiber/matrix interface debonds at different locations in the composite including the composite surface region where the stress state differs from the one in the bulk composite. In the model individual partially debonded fiber is surrounded by matrix region and embedded in a homogenized composite.

1. Introduction

Many engineering structures made from polymer composites such as wind turbine blades, aircraft wings, etc., are subjected to thermal and mechanical fatigue loads. Due to overload or initial manufacturing defects, microdamage such as fiber breaks or matrix cracks may occur after certain amount of load cycles. The presence of microdamage not necessarily means that the structure is not suitable for further service. However, due to further cyclic loading the propagation of microdamage will occur and after a certain number of applied load cycles the mechanical properties will be significantly degraded leading to failure of the structure.

In polymer composites microdamage in form of fiber breaks or matrix cracks is the most common. Under small-medium stress tension-tension fatigue of a UD polymer composite matrix cracks may occur after some number of load cycles. In further fatigue loading the matrix cracks are deflected and the propagation of damage is in form of growing fiber/matrix interface debond cracks. In high stress mechanical fatigue loading, fiber breaks occur in random positions after few loading cycles. Due to energy released during the breaking of the fibers, partial fiber/matrix interface debonding typically occurs in vicinity of fiber breaks. Also in this case the further propagation of damage will be in form

[1] To whom any correspondence should be addressed.

of growing interface debond cracks. In this paper only the latter scenario of damage initiation and propagation is studied.
Characterizing the growth of damage propagation is important to estimate the lifetime and serviceability of the structure.
Analytical [1] and numerical [2] models have been earlier developed to characterize the growth of debond crack when the broken and partially debonded fiber is in the bulk of the composite.
The main objective of the present paper is to analyze the fiber/matrix debond growth as dependent on the fiber position with respect to the specimen surface.

2. Mechanics of debond propagation

Analyzing cyclic loading Paris law, well known for single crack growth in metals, may be suitable for characterization of the debond growth along fiber/matrix interface as a function of applied number of load cycles N. Paris law expression describes the increase of the debond surface area dA as:

$$\frac{dA}{dN} = B(\Delta G_{II})^{m} \tag{1}$$

where: B and m are unknown parameters, which have to be determined experimentally; ΔG_{II} is the energy release rate difference between values corresponding to $\varepsilon_{\max}$ and $\varepsilon_{\min}$ in a cycle. The cyclic loading type considered in this study is mechanical tension-tension fatigue ($\varepsilon_{\max} > 0$ and $\varepsilon_{\min} > 0$).
Notation ΔG_{II} with index II is used since only Mode II debond crack propagation is relevant - for the studied polymer composites the matrix Poisson's ratio is higher than for the fibers and therefore in purely mechanical tension loading the radial stresses on fiber surface will always be compressive, the interface crack is closed and pure Mode II crack propagation will occur.
Debond length increment for circular fiber is related to debond surface increment as $dl_d = dA \cdot (2\pi \cdot r_f)^{-1}$.
Interface debond growth related energy release rate G_{II} can be calculated by various analytical and numerical methods as described in Section 3 of this paper.
The values of two parameters in Paris law (B and m in (1)), however, can only be determined experimentally. Having a reliable model for energy release rate G_{II} calculation in hand, and performing experimental measurements of debond length as a function of applied load cycles N would allow finding the values of Paris law parameters B and m from the best fit.
However, performing experimental measurements of fiber/matrix interface debond crack growth in the bulk of UD composite is a challenging task. Still it could be done using, for example, microtomography techniques. Since Paris law parameters for interface debond growth can be considered as fiber/matrix interface property, a more convenient method would be measuring the debond crack growth in a single fiber composite. Observing single fiber embedded in a transparent block of resin is convenient and can be performed in an optical microscope. In [3,4], for example, debond crack growth in single fiber composites was measured during quasi-static loading. Other option would be to measure debond growth on a composite specimen surface, as, for example, in [5], which can be done in an optical microscope as well. However, the debond growth in vicinity of UD composite surface is different than in the bulk of composite due to edge effects. The high scatter in the experimental measurements in [5] could be an indication that the interface debond growth is very sensitive to the actual distance from the observed fiber to the surface of composite. Therefore, one of the aims of this paper is to calculate the dependency of energy release rate G_{II} on the distance from fiber to composite surface. The change in energy release rate G_{II} is expected due to change in the stress state around the partially debonded fiber when it is in vicinity of composite surface.
It is expected that the stress state around debonded fiber will have higher deviation from axisymmetric stress state, which we have when the fiber is in the bulk of composite, when it is closer to the surface

of composite. In such case 3-D numerical models have to be used for calculation of energy release rate G_{II}.

However, at a certain distance from the composite surface the deviation from axisymmetric stress state will become negligible and more convenient axisymmetric models developed in [1] and [2] can be used for G_{II} calculation.

3. Strain energy release rate for long and short debonds

The debond crack propagation can be considered as self-similar when the tip of the fiber/matrix debond crack is far away from the fiber break where it was initiated and also far from another debond which may be approaching from the other end of the fiber. If the fiber is also sufficiently far from the surface of composite, the stress state around the fiber is axisymmetric and analytical solution for energy release rate G_{II} can be found. In [1] expressions of analytical model for calculation of strain energy are presented.

In case of short debonds, the debond crack tip is close to the fiber break where it initiated from and the stress perturbation regions related to fiber break and debond crack tip interact. Due to interaction, G_{II} is magnified. The shorter is the debond the higher will be the interaction and thus G_{II} will be higher. Since the stress states interact, the self-similar conditions are no longer in power and analytical closed form solution is not possible. Therefore finite element method FEM calculations are useful for determination of G_{II} for short debond lengths. As an output from FEM calculations stress and displacement distributions are available and G_{II} can then be easily calculated, for example, using the virtual crack closure technique [6] as described further. Virtual crack closure technique was used, for example, in [2] to calculate magnification of G_{II} at short debond lengths.

4. Material properties

In this paper two different UD composites were studied – carbon fiber/epoxy resin composite (CF/EP) and glass fiber/epoxy resin composite (GF/EP). The properties of composite constituents are given in Table 1. The accuracy of the values listed in Table 1 are important for correct calculations of the elastic properties of the composite. Importance of this accuracy is discussed in [1].

Table 1. Elastic properties of constituents.

Material	E_L [GPa]	E_T [GPa]	G_{LT} [GPa]	ν_{12} [-]	ν_{23} [-]
CF	500	30	20.00	0.2	0.45
GF	70	70	29.20	0.2	0.20
EP	3	3	1.07	0.4	0.40

Properties of CF/EP and GF/EP unidirectional composites were calculated using the Concentric Cylinder Assembly (CCA) model introduced by Hashin [7,8] for the bonded case. The transverse shear modulus was obtained using the Christensen's self consistent model [9]. Only fiber volume fraction equal to $V_f = 0.5$ was studied in this paper. Fiber radius for all cases was equal to $r_f = 4$ μm.

5. FEM model

5.1. Geometry of the model

For analysis of debond growth in the bulk of composite, axisymmetric FEM models have been developed previously [2]. However, for the case when the fiber is in vicinity of composite surface, axisymmetric model may not be applicable due to edge (surface) effects.

In this study a 3-D FEM model consisting of partially debonded fiber surrounded by a matrix cylinder and embedded in a large block of effective composite was generated using finite element code ANSYS [10]. The geometry of the 3-D FEM model is shown in figure 1. As it is shown in figure 1 only one half of the model was generated in FEM code ANSYS, taking advantage of the symmetry conditions.

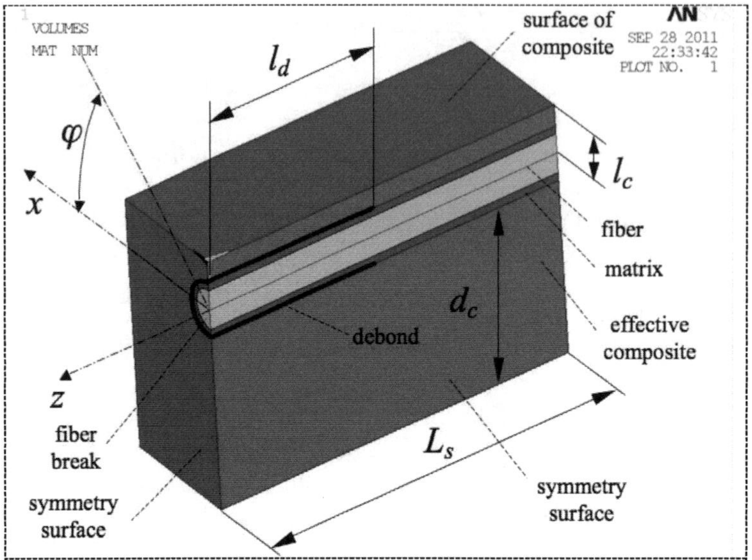

Figure 1. Geometry fragment of the 3-D FEM model.

In figure 1 z, x are the axes in Cartesian system and φ is polar angle, l_d is the variable interface debond crack length, L_s is the total length of model (½ of fiber fragment), l_c is the variable coordinate of depth of the fiber (distance from fiber center axis to composite surface), d_c is the size of the effective composite surrounding the fiber and matrix in the y-direction.

The values of geometrical parameters L_s and d_c were taken from convergence analysis performed in [2] where it was found that model length $L_s = 90 \cdot r_f$ (where r_f is the fiber radius) and the effective composite size $d_c = 5 \cdot r_f$ are efficient for modeling accuracy to represent a long partially debonded fiber in an infinite composite.

Contact elements were used on the surfaces of fiber and matrix in the debonded region. The effect of friction was neglected.

The model was meshed with mesh refinements near the tip of the debond crack in order to obtain accurate solution of stress and displacement distributions which are needed for G_{II} calculations.

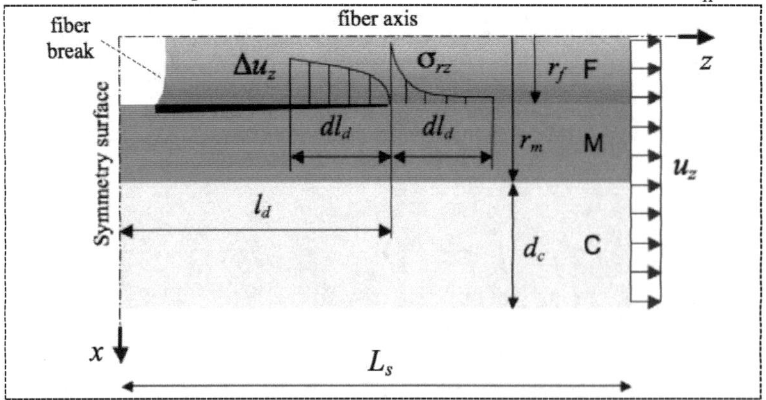

Figure 2. Detail of the FEM model used for G_{II} calculations. F – fiber, M – matrix, C – effective composite.

IOP Conf. Series: Materials Science and Engineering **31** (2012) 012001 doi:10.1088/1757-899X/31/1/012001

Since the virtual crack closure technique was used for G_{II} calculations, the finite element mesh refinement was optimized for the chosen integration length. Figure 2 shows the detail of the FEM model, including the length of integration dl_d.
Following the convergence analysis performed in [2] the length of integration in all calculations was set equal to $dl_d = 1 \cdot r_f$.

5.2. Virtual crack closure technique

The crack closure technique states that the energy released due to debond crack growth by dA is equal to the work which is required to close the newly created surface from size $A + dA$ back to size A. Figure 2 can be used for geometric representation of the further statements.
Closing the debond crack by length dl_d (from $l_d + dl_d$ to l_d) by applying tangential tractions, points at the debonded surface in the region $z \in [l_d; l_d + dl_d]$ which have relative tangential displacement

$$\Delta u^{l_d+dl_d}(z) = u_{fz}^{l_d+dl_d}(z) - u_{mz}^{l_d+dl_d}(z) \tag{2}$$

are moved back to coinciding positions. Here and in following the upper index for stress and displacement shows the length of the debond used in calculations.
At the end of this procedure the shear stress in point z is equal to $\sigma_{xz}^{l_d}(z)$, which is the shear stress in front of the crack with size l_d. For the case when fiber is in the bulk of composite and the stress state around it is axisymmetric, the work required to close the crack by dl_d can be expressed as:

$$dW(dl_d) = 2\pi r_f \frac{1}{2} \int_{l_d}^{l_d+dl_d} \Delta u^{l_d+dl_d}(z) \sigma_{xz}^{l_d}(z) dz \tag{3}$$

In (3) σ_{xz} is the shear stress component relevant to Mode II crack propagation.
For a general 3-D stress state, which may not be axisymmetric the expression (3) can be rewritten as:

$$dW(dl_d) = \frac{1}{2} \int_0^{2\pi} \int_{l_d}^{l_d+dl_d} \Delta u_z^{l_d+dl_d}(z) \sigma_{xz}^{l_d}(z) r_f \, d\varphi dz \tag{4}$$

Within the virtual crack closure technique it is assumed that due to small value of dl_d the relative sliding displacement at the tip of the crack with size $l_d + dl_d$ is the same as at the tip of the debond crack with size l_d:

$$\Delta u^{l_d+dl_d}(z) = \Delta u^{l_d}(z - dl_d) \tag{5}$$

The usefulness of this assumption is that only one stress state calculation for a given debond length is required.
The energy release rate is defined as:

$$G_{II} = \frac{dW}{dA} \tag{6}$$

where $dA = 2\pi r_f dl_d$. Thus G_{II} can be easily calculated from a single stress state by combining equations (4), (5) and (6):

$$G_{II} = \lim_{dl_d \to 0} \frac{1}{4\pi r_f dl_d} \int_0^{2\pi} \int_{l_d}^{l_d+dl_d} \Delta u_z^{l_d+dl_d}(z - dl_d, \varphi)\sigma_{xz}^{l_d}(z,\varphi) r_f \, d\varphi dz \qquad (7)$$

As mentioned, in all calculations performed in this study the length of integration was equal to $dl_d = 1 \cdot r_f$, which as found in [2] can be assumed as sufficiently small to satisfy conditions of (7) (where $dl_d \to 0$) and at the same time sufficiently large to obtain accurate stress and displacement distributions.

Distributions of displacement Δu_z and shear stress σ_{xz} were obtained by performing post-processing (path operations) in ANSYS. Since non-axisymmetric stress state was studied, pre-defined paths were generated along the whole circumference of the fiber/matrix surface.

To facilitate the path operations, the finite element mesh was uniform along the circumference. Code comprising (7) in incremental form was generated for convenience of calculations as an input. Although the stress state around partially debonded fiber which is close to composite surface may not be axisymmetric, the crack growth was considered as uniform – the shape of the interface debond crack front was assumed to be circular with same z coordinates for all points of the tip of the cylindrical crack.

6. Results and discussion

6.1. G_{II} calculations

Figures 3 and 4 show results of interface debond growth related energy release rate G_{II} calculations for CF/EP and GF/EP composites. In all cases uniform axial displacement $u_z = 0.01 \cdot L_s$ was applied at the end surface of the FEM model (see figure 2).

Results in figures 3 and 4 are presented as functions of normalized debond length $l_{dn} = l_d \cdot r_f^{-1}$. The calculations were performed for different cases by varying the fiber depth parameter l_c (distance from composite surface). For CF/EP calculations were done for three cases: l_c = 10, 15 and 20 μm as shown in figure 3. For GF/EP calculations were done for two cases: l_c = 10 and 15 μm as shown in figure 4.

First of all it can be noted from both cases (CF/EP in figure 3 and GF/EP in figure 4) that the values of energy release rate are higher for shorter debond lengths. It is expected that for shorter debond lengths due to interaction of stress states the energy release rate will be magnified. However, when the debond length is sufficiently long, the growth of the debond crack becomes self-similar as it was shown in [2]. In the present paper a range of debond lengths from $l_{dn} = 1.5 \cdot r_f - 4 \cdot r_f$ has been studied. As it can be seen from the trends in figure 3 and figure 4, this range of debond lengths corresponds to interaction region and a further increase of debond length will result in smaller values of G_{II} approaching the asymptotic self-similar value.

Concerning the dependency of G_{II} on fiber distance from surface (parameter l_c) it can be noted for both CF/EP in figure 3 and for GF/EP in figure 4 that G_{II} is higher when the fiber is closer to the surface of composite. Since the energy release rate is higher, it follows that the growth of the interface debond crack will be faster for the fibers which are closer to the composite surface compared to the fibers in the bulk of composite. From calculation results for CF/EP in figure 3 it can be noted that there is no significant difference between results for cases when fiber distance from surface l_c = 15 μm or l_c = 20 μm. This distance l_c = 15 μm, at which energy release rate G_{II} becomes independent of distance from composite surface indicates the boundary between the close-to-surface region and bulk of composite region.

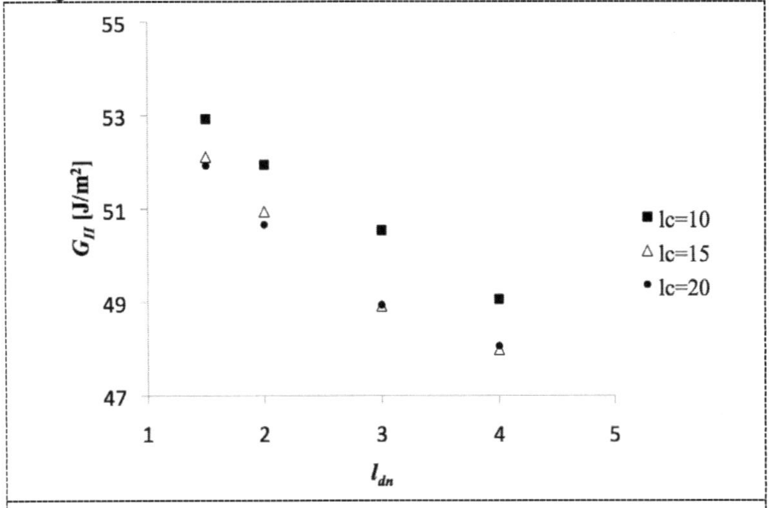

Figure 3. Energy release rate G_{II} as a function of normalized debond length l_{dn} for CF/EP. l_c values are in μm.

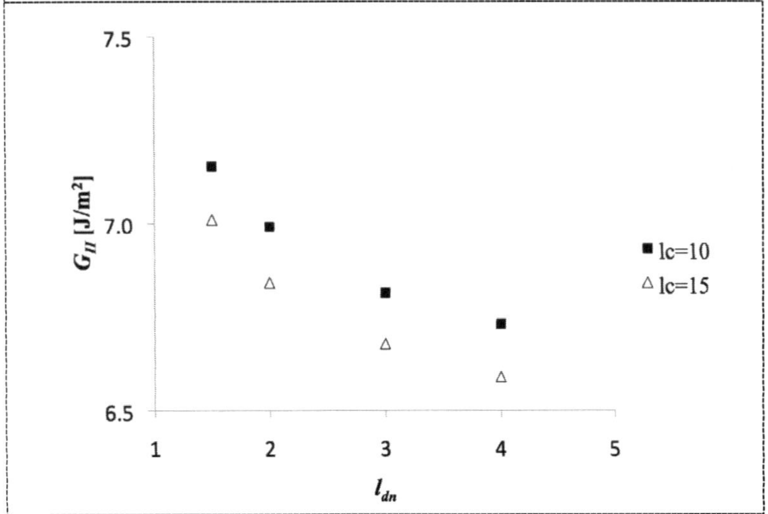

Figure 4. Energy release rate G_{II} as a function of normalized debond length l_{dn} for GF/EP. l_c values are in μm.

To analyze the stress state around the broken and partially debonded fiber and how it depends on the distance from composite surface (l_c) a comparison between the shear stress σ_{xz} and axial displacement Δu distributions was performed.

As it can be seen from results shown in figures 5 and 6 (for CF/EP and GF/EP respectively), where stress distributions are shown for the case when angular coordinate $\varphi = 0°$, there is a minimal dependence on the distance l_c within the observed range from l_c = 10 to 20 μm. In fact, the differences in calculated energy release rate G_{II} values in this range were also relatively small (less than 4%), therefore small differences in stress distributions were expected.

IOP Conf. Series: Materials Science and Engineering **31** (2012) 012001 doi:10.1088/1757-899X/31/1/012001

Regarding the axisymmetry of the stress distribution, dependency of shear stress σ_{xz} distribution on angular coordinate φ was investigated. In figures 7 and 8 stress distributions at 3 different angular directions φ = 0°, 45° and 90° are presented for CF/EP and GF/EP respectively. The results in both figures 7 and 8 indicate that there is no significant dependence of shear stress σ_{xz} distribution on the angular coordinate φ, which means that an axisymmetric approximation is applicable. The same can be concluded from analysis of displacement distributions as shown in figure 9.

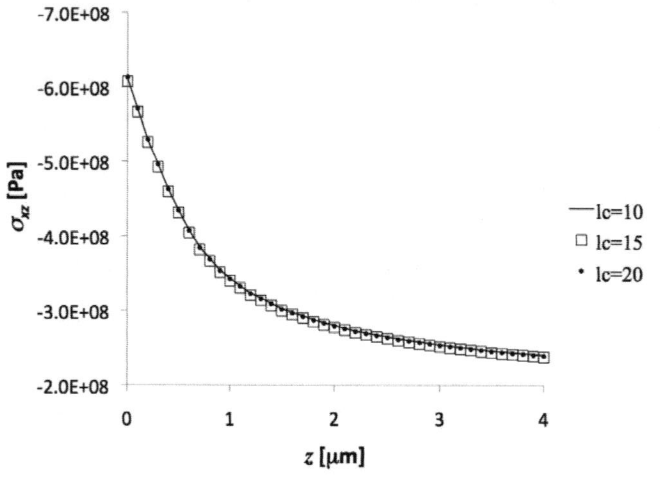

Figure 5. Shear stress σ_{xz} distribution in front of the tip of the debond crack for CF/EP. $z = 0$ is location of the debond crack tip. Angular coordinate in all cases $\varphi = 0°$, l_c values are in μm.

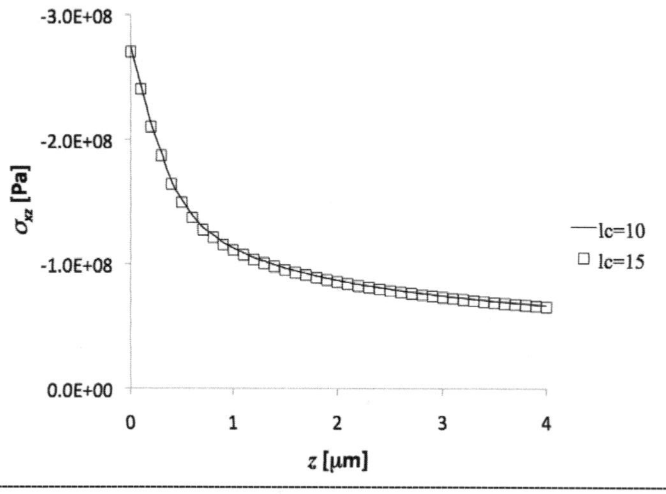

Figure 6. Shear stress σ_{xz} distribution in front of the tip of the debond crack for GF/EP. $z = 0$ is location of the debond crack tip. Angular coordinate in all cases $\varphi = 0°$, l_c values are in μm.

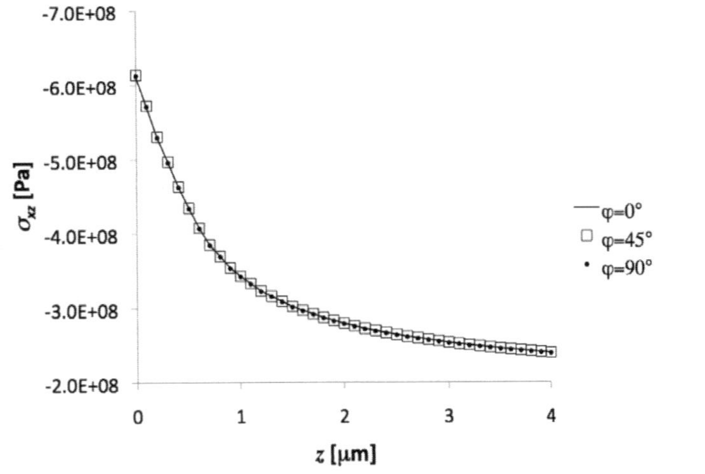

Figure 7. Shear stress σ_{xz} distribution in front of the tip of the debond crack for CF/EP. $z=0$ is location of the debond crack tip. Distance from surface in all cases $l_c = 10$ μm.

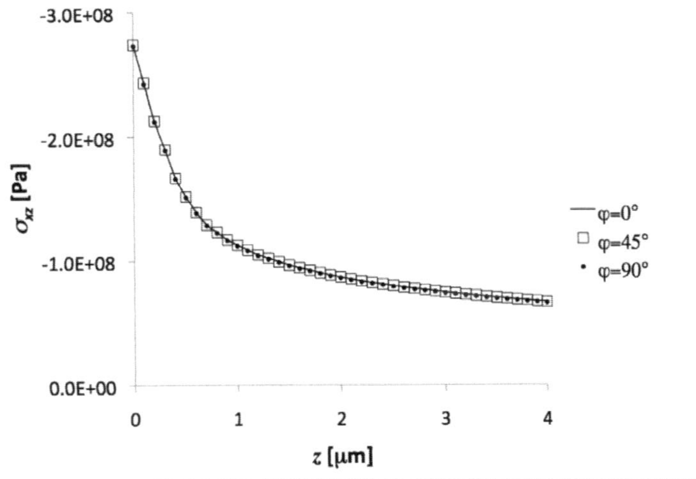

Figure 8. Shear stress σ_{xz} distribution in front of the tip of the debond crack for GF/EP. $z=0$ is location of the debond crack tip. Distance from surface in all cases $l_c = 10$ μm.

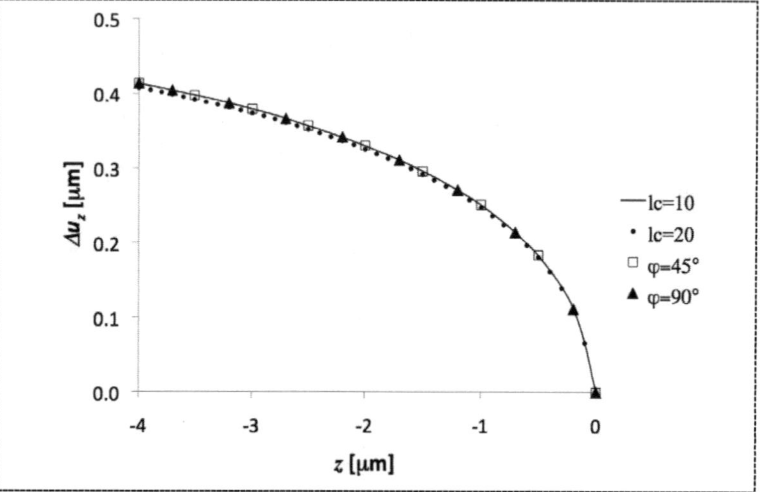

Figure 9. Axial displacement Δu_z distribution behind the tip of the debond crack for CF/EP. $z = 0$ is location of the debond crack tip, l_c values are in μm.

The results show that within the observed region of distances from composite surface, the changes in energy release rate G_{II} values are approximately 1.5-2%.
This means that in the selected range of distances from composite surface l_c, the stress state is very close to the stress state in the bulk of composite. Using the present FEM model geometry (figure 1) the minimal distance from composite surface l_c is limited to approximately $2 \cdot r_f$ and new geometry model with explicit heterogeneous microstructure of the surrounding composite has to be generated to analyze cases when the distance l_c is smaller.

7. Conclusions

The calculation results proved dependency of energy release rate G_{II} on distance from composite surface for both CF/EP and GF/EP composites. The energy release rate is larger for debonds closer to the specimen surface. However, within the observed range (l_c = 10 to 20 μm) the changes in absolute values of G_{II} are small. It was proved that differences due to fiber location in shear stress and axial displacement distributions in front and behind the crack tip were even smaller. Furthermore, dependency of stress distributions on angular coordinate φ was also small, therefore, it can be concluded that axisymmetric approximation can be used in the studied region.

8. References

[1] Pupurs A and Varna J 2011 *Mech. Adv. Mater. Struct.* Accepted
[2] Pupurs A, Krasnikovs A and Varna J 2011 *Mech. Adv. Mater. Struct.* Accepted
[3] Kim B W and Nairn J A 2002 *J. Compos. Mater.* **36** 1825
[4] Kim B W and Nairn J A 2002 *J. Mater. Sci.* **37** 3965
[5] Gamstedt E K 2000 *J. Appl. Polymer Sci.* **76** 457
[6] Irwin G R 1958 *Fracture, Handbuch der Physik* (Berlin: Springer)
[7] Hashin Z and Rosen B W 1964 *J. Appl. Mech.* **31** 223
[8] Hashin Z 1983 *J. Appl. Mech.* **50** 481
[9] Christensen R M and Lo K H 1979 *J. Mech. Phys. Solid.* **27** 315
[10] ANSYS Release 11.0 2007

IOP Conf. Series: Materials Science and Engineering **31** (2012) 012002 doi:10.1088/1757-899X/31/1/012002

Crack opening displacement determination in damaged cross-ply laminate using electronic speckle pattern interferometry (ESPI)

H Zrida [1,2], **M S Loukil**[1,2], **J Varna**[2] and **Z Ayadi**[1]

[1] Institut Jean Lamour, SI2M, EEIGM 6 Rue Bastien Lepage, F-54010 Nancy, France
[2] Division of Polymer Engineering, Luleå University of Technology SE-97187 Luleå, Sweden

E-mails: mohamed-sahbi.loukil@eeigm.inpl-nancy.fr
hana.zrida@eeigm.inpl-nancy.fr

Abstract. Composite laminates during service undergo complex combinations of thermal and mechanical loading leading to microdamage accumulation in the plies. The most common damage mode and the one examined in this work is intralaminar cracking in layers. The crack opening displacement (COD) and the crack sliding displacement (CSD) during loading reduce the average stress in the damaged layer, thus reducing the laminate stiffness.
These parameters depend on material properties of the damaged layer and surrounding layers, on layer orientation and thickness. Previously these parameters have been calculated using finite element method (FEM) assuming linear elastic material with idealized geometry of cracks. To validate these assumptions experimentally the displacement field on the surface of a [90/0/90] carbon fiber/epoxy laminate specimens with multiple intralaminar cracks in the surface layer is studied and the COD dependence on the applied mechanical load is measured. The specimen full-field displacement measurement is carried out using ESPI (Electronic Speckle Pattern Interferometry). The displacement jumps corresponding to cracks are clearly visible and can be used to determine the opening displacement along the cracks. The effect of crack interaction on the COD at high crack density is also investigated.

1. Introduction

When cross-ply laminates are subjected to mechanical loading, different failure mechanisms are induced (see Figure 1). transverse matrix cracking in 90° plies, delamination between 0° and 90° plies, and fiber fracture in 0° plies. In the case of uniaxial loading, the early stage of damage is transverse (intralaminar) matrix cracking in 90° plies. The transverse cracks develop in the fiber direction and extend across the laminates from the free edges of the specimen. The analysis of transverse cracking is important since cracking reduces the stiffness and strength of laminates. Transverse cracks induce local stress concentrations at crack tips and can initiate significant interlaminar delamination between 0° and 90° plies. Fiber fracture in 0° plies is induced only at high loads in the case of monotonic loading or for large cycle number in the case of fatigue loading.

The degradation of the elastic properties of composite laminates is caused by reduced stress in the damaged layer which is mainly due to two parameters: the crack opening displacement (COD) and the crack sliding displacement (CSD). In other words, the elastic modulus in the loading direction and the corresponding Poisson's ratio decrease.

The link between the damaged laminate thermo-elastic properties and the microdamage parameters (COD and CSD) was established in [1,2]. It was shown that only the average values of COD and CSD enter into account in the stiffness expressions directly. In a linear model the average COD and CSD values are proportional to the applied load. Consequently, they have to be normalized to be used in stiffness calculations.

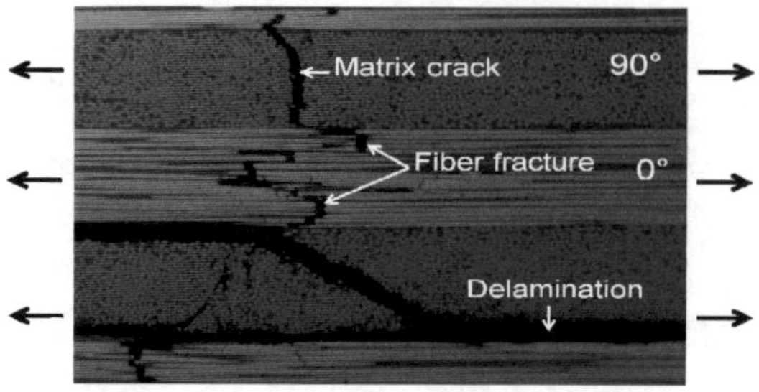

Figure1. Example of damage modes in composites.

FEM studies were performed to understand which material and geometry parameters affect the COD and CSD in [1-3] and simple empirical relationships (power law) were suggested.
The effect of crack density (number of cracks per unit length) on the COD and CSD was analyzed in [3] and studied using FEM in [4]. It was shown that the COD and CSD are smaller for high crack density.

All these studies and analysis assume a linear elastic material with idealized geometry of cracks. The only correct way to validate these assumptions is through experiments.
The effect of material properties on COD was measured experimentally using optical microscopy of loaded damaged specimens in [5,6]. It was shown that the measured average values of COD are affected by the constraining layer orientation and stiffness.

The experimental determination of the average COD and CSD needs the measurement of the displacement for all points of the crack surfaces, which justifies the use of full-field measurement techniques Electronic Speckle Pattern Interferometry (ESPI). ESPI is an optical technique that provides the displacement for every point on a surface and offers the possibility to measure both, the in-plane and out-of-plane displacement without surface preparation.

This technique was used in [7,8] to measure the COD for inside cracks on the specimen's edge. It was shown that the profile of the crack on the edge is elliptical. The main objective of this paper is to study surface cracks by measuring the COD along the crack path. For this reason the cross-ply laminate with surface cracks was selected.
In particular, the displacement field on the surface of a [90/0/90] carbon fiber/epoxy laminate specimen with multiple intralaminar cracks is studied and the relative displacement dependence on the applied mechanical load is measured.

By looking to the displacement field the cracks appear as singularities and the corresponding displacement jumps are directly related to COD and CSD. The transverse cracks are parallel to the fiber orientation in the layer, which in our case corresponds to a 90 direction with respect to the tensile axis (x). Consequently, there is no relative sliding of the crack faces and the only displacement of these crack faces corresponds to COD. In other words, the crack displacement discontinuity measured on the surface is directly the COD.

To investigate edge effects on the opening displacement, we have measured the COD profile as dependent on the distance from specimen edge. The effect of crack interaction on the COD at high crack density is also investigated.

2. Experimental technique and material

2.1. Experimental technique: ESPI

In electronic speckle pattern interferometry, the surface of an object is illuminated by laser light producing a speckled object beam interfering with a coherent reference beam. For in-plane measurement, the two beams are reflected and scattered by the rough measuring surface. The resulting speckle interferogram is imaged on a CCD camera, transferred to a computer, and saved in memory. When the surface is displaced, the speckle interferogram is changed due to the change in path length difference between the object and reference beams. This second interferogram is subtracted pixel by pixel from the first. The result is rectified and displayed as a set of bright and dark fringes, known as correlation fringes, which constitute a contour map of the displacement of the object surface (see Figure 2).

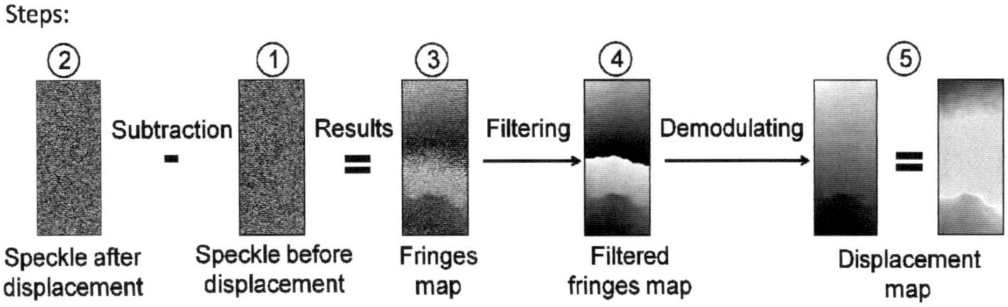

Figure2. Necessary steps to measure the displacement using ESPI.

If the random variation in phase and amplitude of the speckle patterns are not too drastic, the speckle patterns in the original and displaced states are correlated and can be processed to obtain correlation fringes. Excessive displacement can be a source of de-correlation [9,10].

2.2. Material

The sample is a [90,0,90] cross ply laminate made of carbon fiber/epoxy resin.
The elastic properties of the unidirectional composite are the following:
Longitudinal modulus: E_1 (GPa) = 118
Transverse modulus: E_2 (GPa) = 9
Poisson's ratio: $\nu_{12} = 0.3$
In plane shear modulus: G_{12} (GPa) = 4
Prepreg tape thickness: 0.66 mm

A specimen of 13.4mm width and 2mm thickness was reinforced with glass fiber/epoxy resin end tabs in the gripping area. A free length between tabs measures 120mm.

The sample was loaded in tension with displacement rate of 0.5 mm/min in order to introduce damage in the form of micro cracks in the 90-layer. First, the sample was loaded to relatively low strain 0.6% in order to get a low number of cracks to avoid the crack interaction. Afterwards, a higher strain level of loading was applied to introduce more cracks and to study the effect of crack interaction on the COD.

3. Microscopic observation

The cracks in the 90° layer were observed using optical microscopy (figure 3a). They appear on the surface along the fiber direction, hence the name "surface cracks". The top view of one of the surface is schematically presented in Figure 3b and 3c. It is worth noting that no delamination was observed.

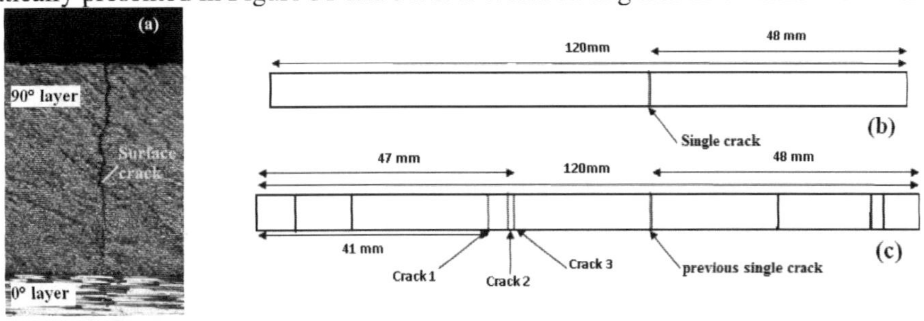

Figure3. Microscopic observation of a surface crack (a).Top view representation of one of the surfaces (b) After low loading: low cracks density = 0.0083 cracks/mm (c) After higher loading: high cracks density = 0.075 cracks/mm.

The geometry of a single crack is presented more clearly in Figure 4.

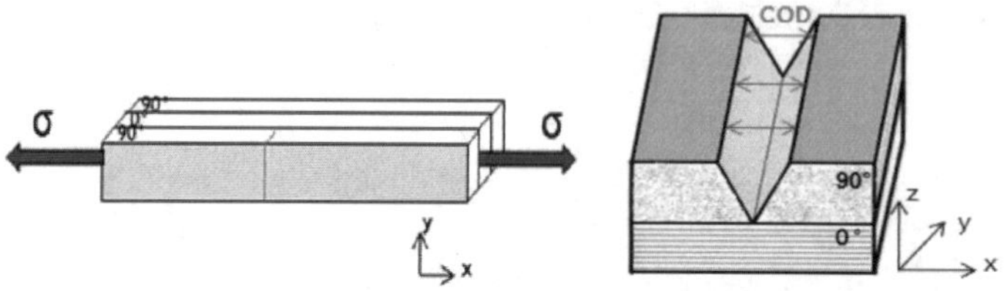

Figure 4. Geometry of a single crack in a sample subjected to tensile loading.

4. COD measurement on a surface crack using ESPI

The experimental arrangement is presented in Figure 5. One end of the sample was clamped and the other one was loaded in tension. The surface was illuminated by a coherent and symmetric laser beam. The observation was made on the surface where the crack has appeared (see Figure 6). The digitization of the 10.71mm x 10.71mm image was performed at 512 x 512 pixels, thus, the spatial resolution is 20.9µm/pixel.

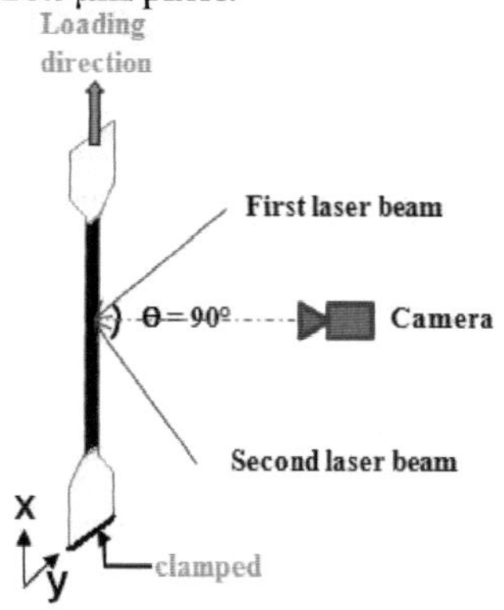

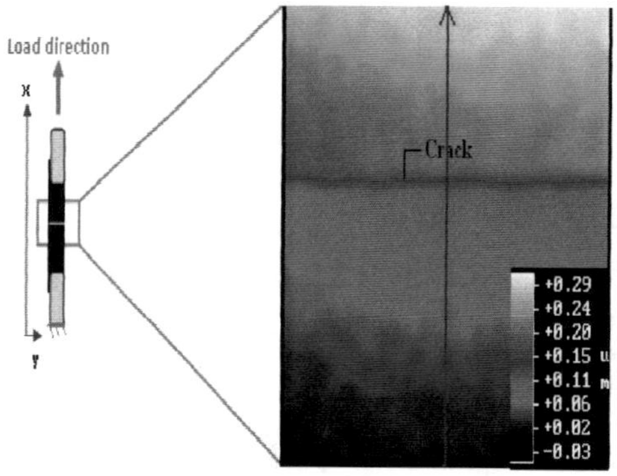

Figure5. The experimental arrangement.

Figure6. Displacement map and crack observation using ESPI

Using the displacement map, the x-displacement (u_x) profile was determined along the X-axis direction (which is the tensile axis) as shown in Figure 7.The displacement discontinuity (jump) caused by the crack is clearly visible.

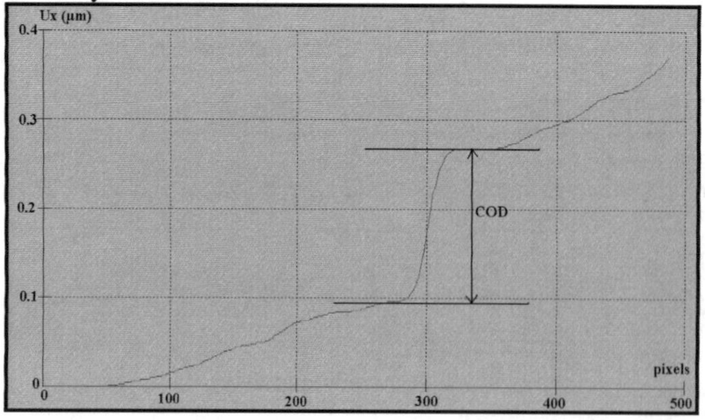

Figure7. Profile of the u_x displacement on the specimen surface (Y=5.226mm) along the tensile axis corresponding to a variation of the relative average stress (Δσ=1,7MPa).

In order to study the evolution of the crack opening displacement across the width of the sample, COD was measured along different vertical lines in the X-axis direction, corresponding to different Y values (width coordinate), see (Figure 8).

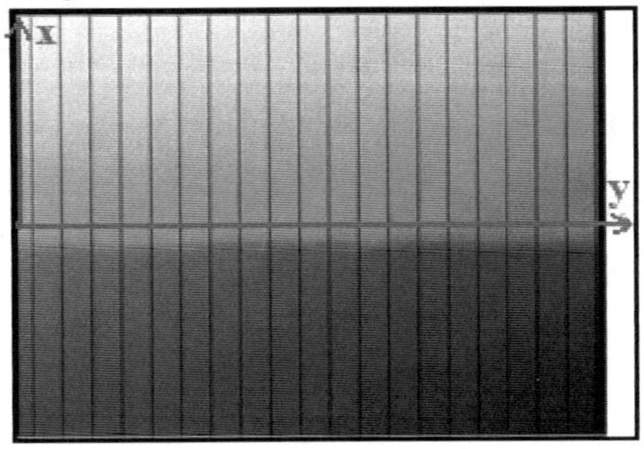

Figure8. Displacement map: u_x measurement.

Displacement maps obtained after a displacement computation process making three times an average of the phase difference using an [11 x 11] filter kernel were used.

The choice of three filtering processes was not arbitrary. The filtering smoothen the displacement curves thus improving the measurement resolution of the displacement but reducing the spatial resolution. Many ways of filtering were tried with the intention to get the best filtering which can provide us with the most accurate results.

Figure 9 shows the displacement u_x measured along the tensile axis on the sample's surface. After the first filtering the displacement slope is close to zero near to the crack surface area where stress free boundary condition is assumed at the micro crack surfaces. These plateau regions become less distinct and virtually disappear with increasing filtering. The COD is measured as the displacement between two crack surfaces where the strain is almost zero (the plateau regions according to Figure 7). Hence, if the plateau regions disappear, we will lose accuracy on this measurement, even if the measurement resolution is much better.

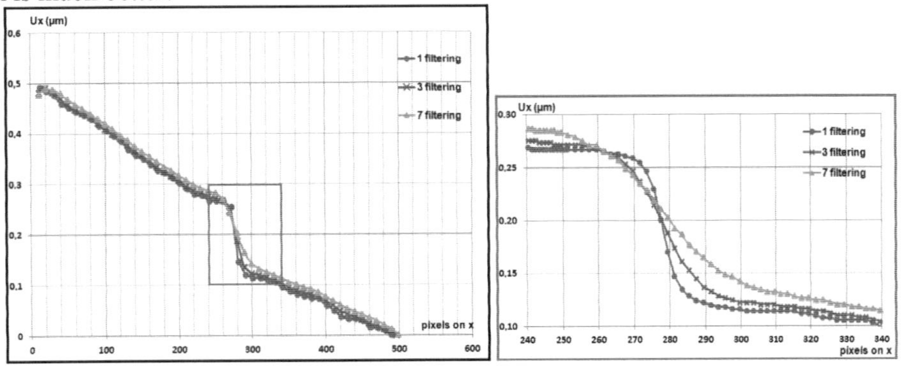

Figure9. Displacement u_x profile for different filtering numbers ($\Delta\sigma$=1,22MPa from σ=37.23 MPa to σ=38.44 MPa).

We have chosen to continue this study with three filtering in an attempt to strike a balance between the two resolutions. Therefore, preserving the plateau regions, we try to improve a little bit the measurement resolution.

5. Results

5.1. Non interactive cracks (single crack)

Figure 10 shows the crack opening displacement as a function of the width coordinate of the sample. The CODs were measured using a displacement map corresponding to the applied average stress change $\Delta\sigma$ =0,73MPa (from σ=37.51 MPa to σ=38.24 MPa). The curve shows that in average the crack opening displacement on the sample surface is about 0.062µm. One can notice the variation of the COD ($\pm$ 14nm) and the trend that the COD is increasing going from the left to the right which could be related to imperfect alignment of the sample in the testing machine. However, there are no noticeable edge effects that would exhibit themselves as larger COD at edges.

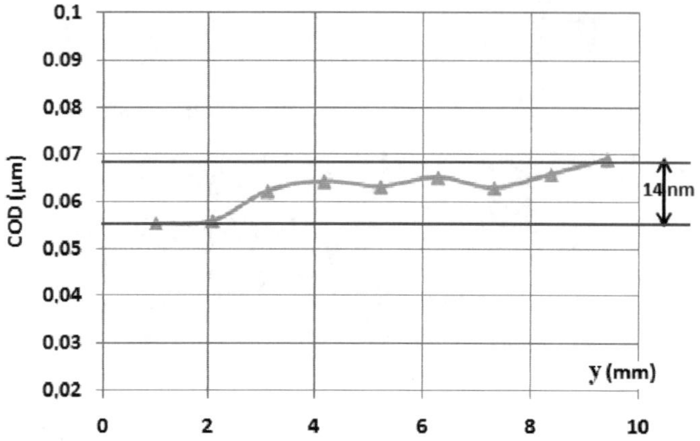

Figure10. COD along the width coordinate for non-interactive crack.

Other tests were carried out with different relative averaged stress $\Delta\sigma$. Figure 11 shows the evolution of the COD (µm) as a function of the averaged applied stress $\Delta\sigma$ (MPa). The COD represented in this figure was determined on a distance Y = 5.226 mm from the edge of the field observed. This curve shows the linear elastic behavior of the composite deduced from the linear variation of the COD with relative average applied stress $\Delta\sigma$. In Figure 12, we notice that the COD increases with increasing loading, but the local variation remains the same whatever the magnitude of the load applied.

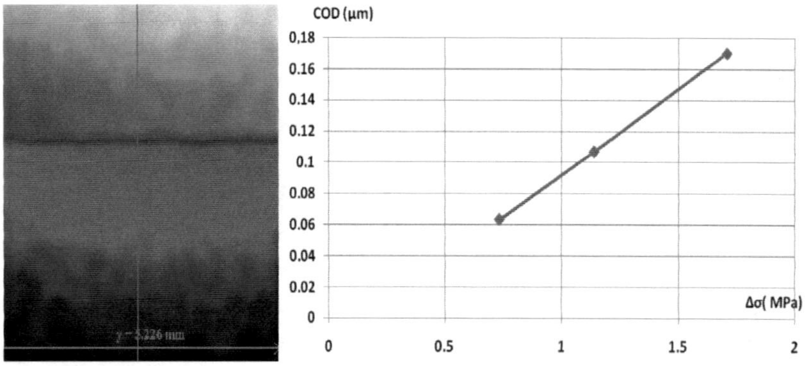

Figure11. COD (µm) as a function of the applied stress Δσ (MPa) in y = 5.226 mm.

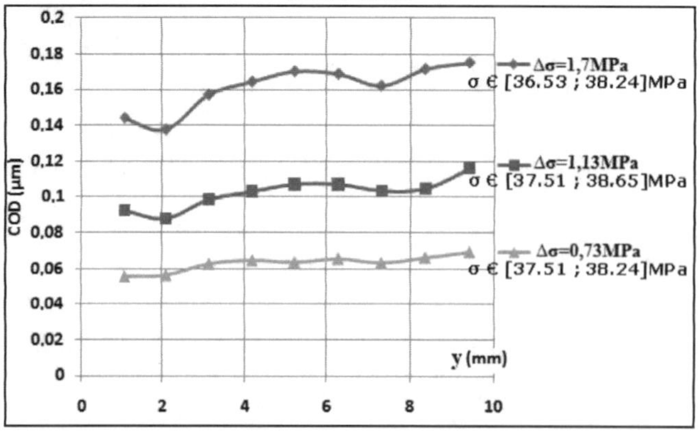

Figure12. COD along the sample's width for different $\Delta\sigma$.

In order to study the edge effect more accurately, the COD on the edges of the sample's width has to be measured. The problem faced in this experiment is that the field observed using ESPI is 10,71mm x 10,71mm. The sample's width is 13,4mm. Knowing that the camera was focused on the middle of the observed surface, there is 1,34mm from the right and the left edges that cannot be seen in the displacement field.

5.2. Interactive cracks

In this section, the focus is on the field where three cracks have emerged after applying high loading (Figure 13). Crack 1 and crack 2 are distanced by 5mm; crack 2 and crack 3 are distanced by 2mm. Since the 90-ply thickness (crack size) is about 0.66 mm these two cracks may be interacting more, which would lead to lower COD's of these cracks.

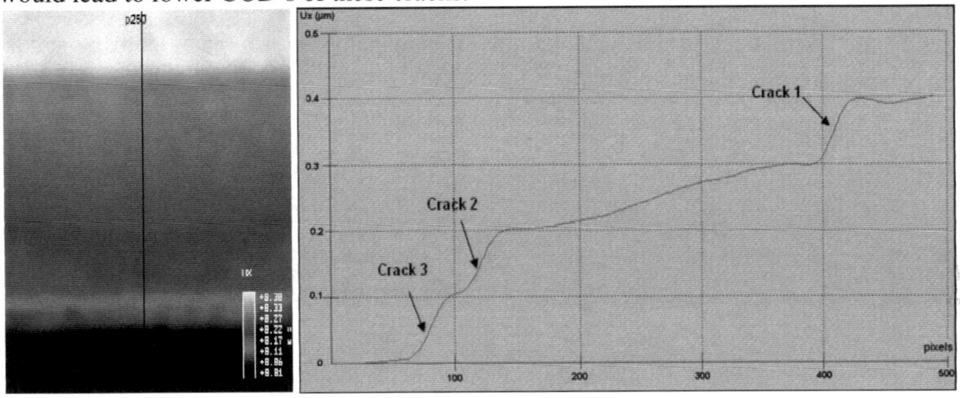

Figure13. Displacement u_x profile along the tensile axis on the surface where 3 cracks exist in the same field (Y = 5.226 mm) (Δσ=0,97 MPa from σ=39,14 MPa to σ=40,11 MPa).

Figure 14 shows COD of each crack along the width. The openings keep almost the same value along the width. A comparison between those three cracks, demonstrates the effect of the crack interaction: crack2 is influenced by two neighboring cracks, that's why the COD2 is smaller than COD1 and COD3 which correspond to crack1 and crack3. When the crack density is high, the stress distributions of two neighboring cracks interact and the average stress between cracks at the given applied load is lower. For this reason, the CODs of the interacting cracks are smaller than those of non-interactive cracks.

If we compare the normalized values with the normalized value for the single crack analyzed in Section 5.1, we find that the CODs of interacting cracks are significantly smaller than the CODs of non-interacting cracks. COD = 0,085µm/MPa for single crack and COD = 0,059µm/MPa for interactive cracks. Certainly this result should be considered with a caution because due to statistical variation of the micro-heterogeneity of the composite there may be a large variation in crack opening of cracks at different locations.

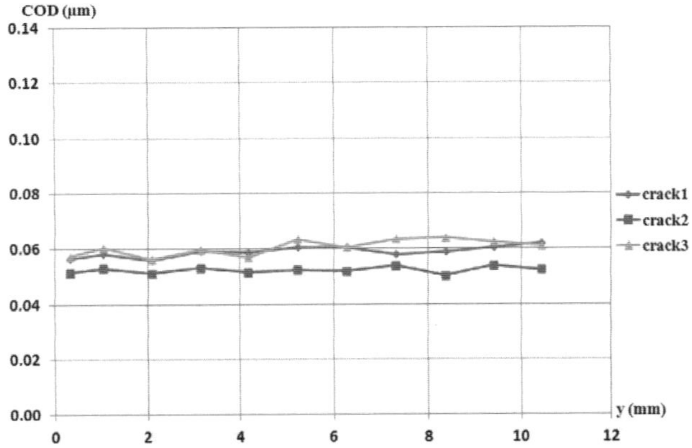

Figure14. COD along the sample's width for interactive cracks (Δσ=0,97 MPa from σ=39,14 MPa to σ=40,11 MPa).

6. Conclusions

Using ESPI, the displacement field on the surface of a [90/0/90] carbon fiber/epoxy laminate specimens with multiple intralaminar cracks was studied and the crack opening displacement, COD dependence on the applied mechanical load was measured.

The presented study proves the potential of the ESPI allowing measuring of the crack opening displacement along the specimen width to investigate the edge effect. The opening of surface cracks is uniform along the sample width for non-interactive cracks as well as for interactive cracks not showing any edge effects. It was confirmed that the COD is proportional to the applied load. The crack interaction at high crack density leads to significant decrease of the normalized COD.

7. References

[1] Lundmark P and Varna J 2005 Constitutive relationships for laminates with ply cracks in in plane loading *Int. J. of Damage Mech.* **14 (3)** 235-61
[2] Lundmark P and Varna J 2006 Crack face sliding effect on stiffness of laminates with ply cracks *Comp. Sc. Tech.***66** 1444–54
[3] Joffe R, Krasnikovs A and Varna J 2001 COD-based simulation of transverse cracking and stiffness reduction in [S/90n]s laminates *Comp. Sc. and Tech.* **61** 637-56
[4] Lundmark P, Varna J 2011 Stiffness reduction in laminates at high intralaminar crack density: effect of crack interaction *Int. J. of Damage Mech.* **20** 279-97
[5] Varna J, Berglund LA, Talreja R and Jakovics A 1993 A study of the crack opening displacement of transverse cracks in cross ply laminates *Int. J. of Damage Mech.* **2** 272–89
[6] Varna J, Joffe R, Akshantala NV and Talreja R 1999 Damage in composite laminates with off-axis plies *Comp. Sc. Tech.* **59** 2139–47
[7] Farge L, Ayadi Z and Varna J 2008 Optically measured full-field displacements on the edge of a cracked composite laminate *Comp. Part A* **39** 1245-52
[8] Farge L, Varna J and Ayadi Z 2010 Damage characterization of a cross-ply carbon fiber/epoxy laminate by an optical measurement of the displacement field *Comp. Sc. and Tech.* **70**, 94-101
[9] Moore AJ and Tyrer JR 1996 Two-dimensional strain measurement with ESPI *Opt. Laser Eng.* **24** 381-402.
[10] Nakadate S and Saito H 1985 Fringe scanning speckle interferometry *Appl. Opt.* **24(14)** 2172-80

IOP Conf. Series: Materials Science and Engineering **31** (2012) 012003 doi:10.1088/1757-899X/31/1/012003

Minimization of complementary energy to predict shear modulus of laminates with intralaminar cracks

K Giannadakis and J Varna[1]
Department of Engineering Sciences and Mathematics, Division of Materials Science, Lulea University of Technology, SE-97187, Lulea, Sweden

E-mail: Konstantinos.Giannadakis@ltu.se, Janis.Varna@ltu.se

Abstract. The most common damage mode and the one examined in this work is the formation of intralaminar cracks in layers of laminates. These cracks can occur when the composite structure is subjected to mechanical and/or thermal loading and eventually lead to degradation of thermo-elastic properties. In the present work, the shear modulus reduction due to cracking is studied. Mathematical models exist in literature for the simple case of cross-ply laminates. The in-plane shear modulus of a damaged laminate is only considered in a few studies. In the current work, the shear modulus reduction in cross-plies will be analysed based on the principle of minimization of complementary energy. Hashin investigated the in-plane shear modulus reduction of cross-ply laminates with cracks in inside 90-layer using this variational approach and assuming that the in-plane shear stress in layers does not depend on the thickness coordinate. In the present study, a more detailed and accurate approach for stress estimation is followed using shape functions for this dependence with parameters obtained by minimization. The results for complementary energy are then compared with the respective from literature and finally an expression for shear modulus degradation is derived.

1. Introduction

In laminated composite materials, many different kinds of micro-damage modes may evolve without leading to final failure, causing however stiffness reduction. One of the first and the most common damage mode in laminates causing degradation of thermo-elastic properties is intralaminar cracking. Whereas numerous approaches and analytical models have been developed to study the reduction of the elastic modulus and poisson's ratio, the in-plane shear modulus of the damaged laminate is considered only in a few investigations.

Hashin [1] developed a model for the in-plane shear modulus reduction of cross-ply laminates with cracks in central 90-layer using a principle of minimum complementary energy. According to this principle the most accurate, between all "admissible" stress states, is the one which gives the lowest value of the complementary energy. An admissible stress field is a constructed stress state which satisfies equilibrium as well as the boundary and interface conditions in tractions. The in-plane shear stress distribution across each ply was assumed uniform and the only unknown function describing the in-plane shear stress distribution along the specimen axis was determined using the principle of

[1] Corresponding author: Janis.Varna@ltu.se

minimum complementary energy. Using the calculated stress distribution the lower bounds for the shear modulus of the laminate was obtained.
The same solution for shear modulus of a cross-ply laminate with cracks in 90-layer was obtained in [2,3] using different approaches. Tan et al. [2] obtained expressions for shear modulus integrating the equilibrium and constitutive expressions over the ply thickness. Abdelrahman et al [3] reduced the 3-D elasticity problem to 2-D problem in terms of stresses averaged over the ply thickness. In both cases differential equations with constant coefficients for stress distribution were obtained and solved analytically. The results of all these models coincide due to assumed linear through the thickness dependence of the out-of-plane shear stresses.
Fan et al [4] presented constitutive equations for a layer with cracks. These expressions contain the so-called "in-situ damage effective functions" dependent on crack density in the lamina and on the neighbouring layer constraint. In order to determine these functions "an equivalent constraint model" was introduced, assuming that the constraint of the lay-ups above and below the damaged lamina can be replaced by two sublaminates with properties calculated using laminate theory (CLT). Thereby the actual laminate was replaced by a cross-ply with orthotropic constraint layers. The stress state was calculated using shear lag model with linear distribution of out-of-plane shear stresses. Since the model coincides with the presented in [5], the predictions for cross-ply laminate with cracks in 90-layer only also coincide or are very similar. This approach was refined in Zhang et al in [6] solving the local stress problem using shear lag model which assumes that intralaminar shear stress in 0-layer is present only in a part of the constraint layer. Unfortunately, the size of this zone becomes a fitting parameter.
Tsai et al [7] and also Kashtalyan et al in [8] considered shear response of laminates with cracks in several layers. In this case the set of 2-D equations was solved numerically using finite differences. They suggested that an expression based on "superposition of solutions" may give good accuracy. This problem is out of the scope of the present paper.
Gudmundson et al. [9] and Varna and Lundmark [10] considered damaged laminates with general lay-up and used homogenization technique to derive expressions for stiffness and thermal expansion coefficient tensors. These expressions in an exact form correlate damaged laminate thermo-elastic properties with parameters characterizing crack behavior: crack densities, the average crack opening and crack face sliding. Gudmundson suggested to neglect the effect of neighboring layers on crack face displacements and to determine them using the known solution for a periodic system of cracks in an infinite homogeneous transversely isotropic medium (90-layer). In [10] the crack face sliding was analyzed using FEM and found that it depends on the constraining ply and cracked ply shear modulus ratio and thickness ratio. This relationship was approximated by a power law.
There is an obvious lack of experimental data regarding shear modulus reduction to validate these models. The origin of the experimental difficulties is as follows: intralaminar cracks are usually not introduced in shear but in loadings where Mode I is dominant (for example tensile tests, fatigue or thermal cycling) and the geometry of these specimens is not suitable for shear modulus determination after crack introduction. May be the best lay-up to measure shear modulus degradation would be $[45/-45]_s$ but additional difficulties related to nonlinear material behaviour in shear have to be expected. Tsai et al [5] developed a methodology for shear modulus determination using an experimental setup where a pre-damaged tensile specimen is subjected to in-plane tangential displacement in the middle part. Then the shear modulus was calculated using a Timoshenko beam approximation.

The theoretical models described above are either too simple or too complex to understand or to use. Due to insufficient experimental data FEM has to be used for validation. The objective of the presented work is to develop a more accurate model than in [1,2,3,5], keeping the damaged laminate shear modulus expressions in a compact and easy-to use form. The principle of minimum complementary energy will be used to find the stress state of the Representative Volume Element (RVE) between two cracks.

2. Theoretical Modeling

The RVE (a repeating element in a periodic system of cracks) in $[S,90_n]_s$ laminate with cracks in the 90-layer as shown in figures1a-1b is considered. S is the notation used for balanced "sub-laminate" which in the simplest case could be a unidirectional 0-layer. The crack spacing is $2l_0$, 90-layer thickness $2d$, the sub-laminate's thickness is b and the laminate's overall thickness is $2h = 2(b+d)$. The shear modulus of the undamaged laminate, G_{xy0}^{LAM} can be calculated using Classical laminate Theory (CLT) as follows:

$$G_{xy0}^{LAM} = \frac{1}{h}\left(G_{xy}^{S} b + G_{12} d\right) \tag{1}$$

In (1), G_{xy}^{S} is the shear modulus of the sub-laminate which also may be calculated using CLT and G_{12} is the shear modulus of the unidirectional (UD) composite. It is assumed that the laminate is subjected to average in-plane shear stress τ_0. Then the undamaged laminate shear strain, γ_{xy0}^{LAM} and shear stresses in layers, τ_0^{90}, τ_0^{S} are:

$$\gamma_{xy0}^{LAM} = \tau_0 / G_{xy0}^{LAM} \qquad \tau_0^{90} = G_{12}\gamma_{xy0}^{LAM} \quad \tau_0^{S} = G_{xy}^{S}\gamma_{xy0}^{LAM} \tag{2}$$

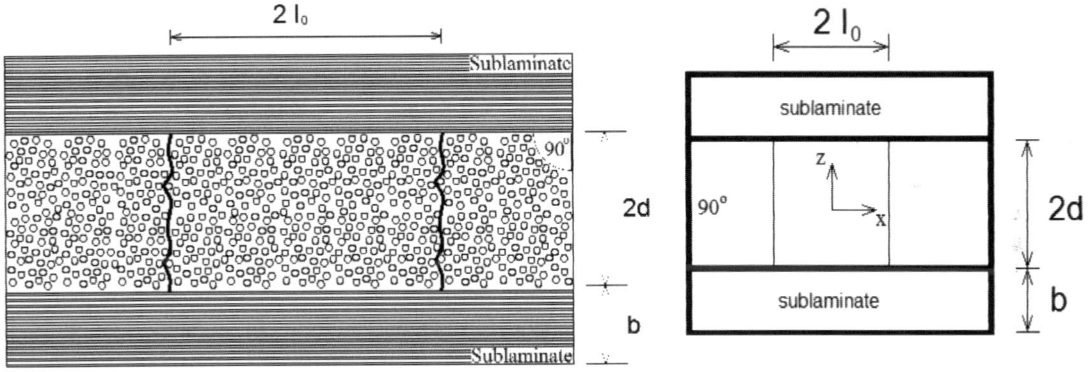

Figure 1: a: Symmetric and balanced laminate with cracks in 90-layers, b: Mathematical analogue

2.1 Assumed stress distributions

The modeling starts with the assumption that the in-plane shear stress in the cracked 90-layer, σ_{xy}^{90}, does not depend on the thickness coordinate z, whereas in the sub-laminate supporting this layer, the z-distribution is given by $f'(z)$. Here and in following, $(\ )'$ denotes derivative of the function of one variable. In the sub-laminate the shear stress is in a form of a simple product of x and z dependent functions which is the main limitation to accuracy. Employing two unknown stress perturbation functions $\Phi(x)$and $\Phi_1(x)$ to describe the shear stress distribution in x-direction we use the following form of the shear stresses:

$$\sigma_{xy}^{90} = \tau_0^{90} \cdot (1 - \Phi(x)) \tag{3}$$

$$\sigma_{xy}^{S} = \tau_0^{S} \cdot (1 - d \cdot \Phi_1(x) \cdot f'(z)) \tag{4}$$

These forms are a generalization of expressions used by Hashin [1] where these stress components were assumed z -independent.
The shear stress in the 90-layer has to satisfy the traction free condition on crack surfaces

$$\sigma_{xy}^{90}(x=\pm l_0)=0 \tag{5}$$

The solution has to satisfy stress free boundary conditions on $z=0;h$ for components σ_{yz}, where $h=b+d$. In addition to that, the equilibrium equations in each point and in average as well as interface conditions in tractions have to be satisfied. The principle of minimum complementary energy will be used next, in order to insure approximate displacement continuity on the interface and in layers.

From the shear force balance

$$\int_0^h \sigma_{xy} dz = \tau_0 (d+b) \tag{6}$$

we obtain

$$\Phi_1(x) = -\Phi(x)\frac{G_{12}}{G_{xy}^S}\frac{1}{f(h)-f(d)} \tag{7}$$

leading to

$$\sigma_{xy}^S = \tau_0^S + \tau_0^{90}\cdot d\cdot \frac{f'(z)}{f(h)-f(d)}\cdot\Phi(x) \tag{8}$$

Substituting σ_{xy}^S in the equilibrium equation

$$\frac{\partial\sigma_{yx}}{\partial x}+\frac{\partial\sigma_{yz}}{\partial z}=0 \tag{9}$$

integrating it with respect to x and applying stress free condition on the surface $z=h$ we obtain

$$\sigma_{yz}^S = \tau_0^{90}\cdot d\cdot\left[\frac{f(h)-f(z)}{f(h)-f(d)}\right]\cdot\Phi'(x) \tag{10}$$

The same equilibrium equation and boundary condition at $z=0$ in 90-layer yields

$$\sigma_{yz}^{90}=\tau_0^{90}\cdot\Phi'(x)\cdot z \tag{11}$$

It can be checked that continuity of σ_{yz} at the layer interface is satisfied. By introducing a new unknown function $\varphi(z)$ as:

$$\varphi = \frac{f(h)-f(z)}{f(h)-f(d)} \tag{12}$$

the following stress expressions can be written:

$$\sigma_{xy}^{90}=\tau_0^{90}\cdot(1-\Phi(x)) \tag{13}$$

$$\sigma_{xy}^S=\tau_0^S-\tau_0^{90}\cdot d\cdot\Phi(x)\cdot\varphi'(z) \tag{14}$$

$$\sigma_{yz}^{90}=\tau_0^{90}\cdot\Phi'(x)\cdot z \tag{15}$$

$$\sigma_{yz}^S=\tau_0^{90}\cdot d\cdot\varphi(z)\cdot\Phi'(x) \tag{16}$$

Finally, the unknown functions $\varphi(z)$ and $\Phi(x)$ have to satisfy boundary conditions

$$\varphi(h)=0 \qquad \varphi(d)=1 \qquad \Phi(\pm l_0)=1 \tag{17}$$

2.2 Minimization of complementary energy

When calculating elastic properties, the thermal stresses are with no importance and therefore thermal terms can be neglected in the complementary energy expression. Therefore, the complementary energy becomes equal to the strain energy. The strain energy of a repeating element between two cracks in 90-layer is

$$U^S = \frac{1}{2}\cdot\int_{-l_0}^{l_0}\int_{h_S}\left[\frac{\left(\sigma_{xy}^S\right)^2}{G_{xy}^S}+\frac{\left(\sigma_{yz}^S\right)^2}{G_{yz}^S}\right]dzdx \tag{18}$$

$$U^{90} = \frac{1}{2}\cdot\int_{-l_0}^{l_0}\int_{h_{90}}\left[\frac{\left(\sigma_{xy}^{90}\right)^2}{G_{12}}+\frac{\left(\sigma_{yz}^{90}\right)^2}{G_{12}}\right]dzdx \tag{19}$$

Symbols $\int_{h_S}$ and $\int_{h_{90}}$ are used for integration over the corresponding layer thickness. Substituting (13)-(16) in (18), (19) and performing z-integration we obtain the following expression for complementary energy of the laminate in case of cracked 90-layer

$$U^{LAM} = U^S + U^{90} = U_0^{LAM} + \frac{\left(\tau_0^{90}\right)^2 d}{2G_{12}}\int_{-l_0}^{+l_0}\left[A_1\cdot\Phi^2(x)+A_2\cdot d^2\cdot[\Phi'(x)]^2\right]dx \tag{20}$$

where

$$A_1 = 1+\frac{G_{12}}{G_{xy}^S}K_1 \qquad A_2 = \frac{1}{3}+\frac{G_{12}}{G_{yz}^S}K_2 \tag{21}$$

$$K_1 = d\int_d^h[\varphi'(z)]^2dz \qquad K_2 = \frac{1}{d}\int_d^h[\varphi(z)]^2dz \tag{22}$$

The function Φ that minimizes (20) has to satisfy the Euler-Lagrange equation

$$\frac{d}{dx}\left(\frac{\partial L}{\partial\Phi'}\right)-\frac{\partial L}{\partial\Phi}=0 \tag{23}$$

which leads to the following equation for $\Phi(x)$ determination

$$d^2\cdot A_2\cdot\Phi''(x)-A_1\cdot\Phi(x)=0 \tag{24}$$

The solution of (24) which satisfies the boundary condition (17) is:

$$\Phi = \frac{\cosh(\mu\frac{x}{d})}{\cosh(\mu\frac{l_0}{d})}, \qquad \text{where } \mu = \sqrt{\frac{A_1}{A_2}} = \sqrt{\frac{1+\frac{G_{12}}{G_{xy}^S}K_1}{\frac{1}{3}+\frac{G_{12}}{G_{yz}^S}K_2}} \tag{25}$$

The solution (25) provides the minimum of the complementary energy for the used selection of $\varphi(z)$ which means that it can be further minimized by proper selection of $\varphi(z)$. For this purpose the complementary energy (20) can be rearranged as follows:

$$U^{LAM} = U_0^{LAM} + \frac{\left(\tau_0^{90}\right)^2 d}{2G_{12}}\left\{\int_{-l_0}^{+l_0}A_1\cdot\Phi^2(x)dx + A_2\cdot d^2\left[\Phi(x)\cdot\Phi'(x)\Big|_{-l_0}^{+l_0}-\int_{-l_0}^{+l_0}\Phi(x)\Phi''(x)dx\right]\right\} \tag{26}$$

Replacing Φ'' in the second integral via Φ as follows from (24) we obtain

$$U^{LAM} = U_0^{LAM} + \frac{(\tau_0^{90})^2 d^3}{2G_{12}} A_2 2\Phi'(l_0) \tag{27}$$

Substituting in (27) A_2 given by (21) and expressing $\Phi'(l_0)$ from (25) we obtain

$$U^{LAM} = U_0^{LAM} + \frac{(\tau_0^{90})^2 d^2 \mu}{G_{12}}\left(\frac{1}{3} + \frac{G_{12}}{G_{yz}^S} K_2\right)\tanh\left(\mu \frac{l_0}{d}\right) \tag{28}$$

Selecting different functions $\varphi(z)$ we will have different values of K_2 and μ, leading to different values of U_{LAM}. The lowest value will correspond to the most accurate stress distribution in average.

2.3Shear modulus expression

The shear modulus of the damaged laminate can be determined using u^{LAM} - the strain energy density of the RVE

$$\gamma_{xy}^{LAM} = \frac{\partial u^{LAM}}{\partial \tau_0} \tag{29}$$

The strain energy density can be obtained by dividing (28) with the RVE volume. Moreover, the shear stress $\tau_0^{(90)}$ in (28) can be expressed through τ_0 using (2) and the undamaged laminate strain energy density is $\tau_0^2 / 2G_{xy0}^{LAM}$. Performing differentiation according to (29) we obtain

$$\gamma_{xy}^{LAM} = \frac{\tau_0}{G_{xy0}^{LAM}} + \frac{\tau_0 G_{12}}{(G_{xy0}^{LAM})^2}\frac{d^2}{(d+b)l_0}\mu\left(\frac{1}{3} + \frac{G_{12}}{G_{yz}^S} K_2\right)\tanh\left(\mu \frac{l_0}{d}\right) \tag{30}$$

Since the Hooke's law for the damaged laminate has to have the form (2), the following shear modulus expression for the damaged laminate can be obtained:

$$\frac{G_{xy}^{LAM}}{G_{xy0}^{LAM}} = \frac{1}{1 + \frac{G_{12}}{G_{xy0}^{LAM}}\frac{d}{d+b}\frac{d}{l_0}\mu\left(\frac{1}{3} + \frac{G_{12}}{G_{yz}^S} K_2\right)\tanh\left(\mu \frac{l_0}{d}\right)} \tag{31}$$

For cross-ply laminates, $G_{xy}^S = G_{12}$, $G_{yz}^S = G_{23}$ and $G_{xy0}^{LAM} = G_{12}$. Then the expressions for the shear modulus can be re-written as:

$$\frac{G_{xy}^{LAM}}{G_{xy0}^{LAM}} = \frac{1}{1 + \frac{d}{d+b}\frac{d}{l_0}\mu\left(\frac{1}{3} + \frac{G_{12}}{G_{23}} K_2\right)\tanh\left(\mu \frac{l_0}{d}\right)}, \qquad \mu = \sqrt{\frac{1+K_1}{\frac{1}{3} + \frac{G_{12}}{G_{23}} K_2}} \tag{32}$$

The shear modulus reduction is usually presented as a function of crack density, ρ (cr/mm) or normalized crack density, ρ_n

$$\rho = \frac{1}{2l_0} \qquad \rho_n = \frac{2d}{2l_0} \tag{33}$$

2.4Selection of $\varphi(z)$ function

Since it is expected that the shear stress perturbation will decay with the increasing distance from the crack tip (layer interface), the absolute values of the $\varphi(z)$ function and its first derivative have to be decreasing with increasing distance from the interface.

A natural choice with these features is an exponential function. The shape function will contain one shape parameter n, which is used to distinguish the minimum of the complementary energy given by (28). For the present study, the function that satisfies boundary conditions (17) is selected in the following form

$$\varphi(z) = \frac{1}{e^{\frac{n \cdot b}{d}} - 1} \cdot \left(e^{\frac{n \cdot (z-h)}{d}} - 1 \right) \quad \text{and} \quad \varphi' = -\frac{n}{d} \cdot \frac{e^{\frac{n \cdot (z-h)}{d}}}{e^{\frac{n \cdot b}{d}} - 1} \tag{34}$$

Using (22) the parameters K_1, K_2 are calculated

$$K_1 = \frac{n}{2} \frac{e^{\frac{nb}{d}} + 1}{e^{\frac{nb}{d}} - 1} \qquad K_2 = \frac{\frac{b}{d} - \frac{1}{2n}\left(1 - e^{\frac{2nb}{d}}\right) + \frac{2}{n}\left(1 - e^{\frac{nb}{d}}\right)}{\left(e^{\frac{nb}{d}} - 1\right)^2} \tag{35}$$

It can be proven that Hashin's solution [1] is a particular case of the selected function (34). In Hashin's model the function $\varphi(z)$ gives linear out-of-plane shear stress distribution in the sublaminate

$$\varphi_{Hashin}(z) = \frac{h - z}{b} \tag{36}$$

Writing the Taylor expansion of (34) for small n values we obtain

$$\lim_{n \to 0} \varphi(z) = \frac{1 - \frac{n \cdot (z-h)}{d} - 1}{1 + \frac{n \cdot b}{d} - 1} = \frac{h - z}{b} = \varphi_{Hashin}(z) \tag{37}$$

Moreover, in Hashin's case

$$K_{1,Hashin} = \frac{d}{b} \qquad K_{2,Hashin} = \frac{b}{3d} \tag{38}$$

3. Results

3.1 Shear modulus evolution

In order to define the parameter n, minimization of (28) was performed. Parameter n was increased with a small step and at each value the perturbation complementary energy (the second term in (28)) was calculated. An example of these calculations for $[0/90]_s$ laminate with $\rho_n = 0.3$ is shown in figure 2a). There is only one minimum in the curve and it is usually in the region $n \notin (0,5)$. The n value which corresponds to the minimum becomes smaller with increasing crack density, see figure 2b. Once the n parameter has been defined for each crack density, the in-plane shear modulus reduction can be calculated according to (31).

For very high density of transverse cracks the shear modulus expression (31) and any other model have to approach to predictions of the ply-discount model which assumes that transverse and shear modulus of the damaged layer is zero. One can show that it is possible only if the shape parameter $n \to 0$. Thus the solution in the very high crack density region coincides with the Hashin's solution. The evolution of the laminate shear modulus as a function of crack density is presented in figure 3. Figures 3a-3c present the cases of a cross-ply laminate with different ply thicknesses and out-of plane elastic constants, whereas in figure 3d, the general case of a sublaminate used as outer layer is presented.

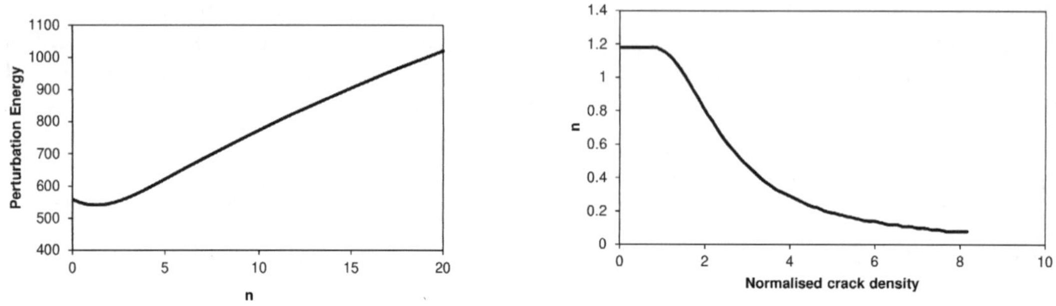

Figure 2 a: Perturbation energy dependence on the shape parameter n showing minimum in small n –value region. b: the "best" n dependence on crack density. Cross-ply laminate with d=b=0.6mm, l_0= 2mm, G_{12}=4.58GPa, G_{23}=3.4GPa

The presented results show that the developed model gives higher values of the residual laminate shear modulus than Hashin's model. It is an agreement with the general feature that the solutions based on minimization of the complementary energy give lower bound to the exact solution and the approximate solutions approach to the exact solution when the accuracy of the stress state is improved. Still, the improvement is rather marginal for laminates with large d/b ratio (figure 3a). The difference between models becomes larger if the cracked layer is supported by sublaminate with large shear modulus, see figure 3d.

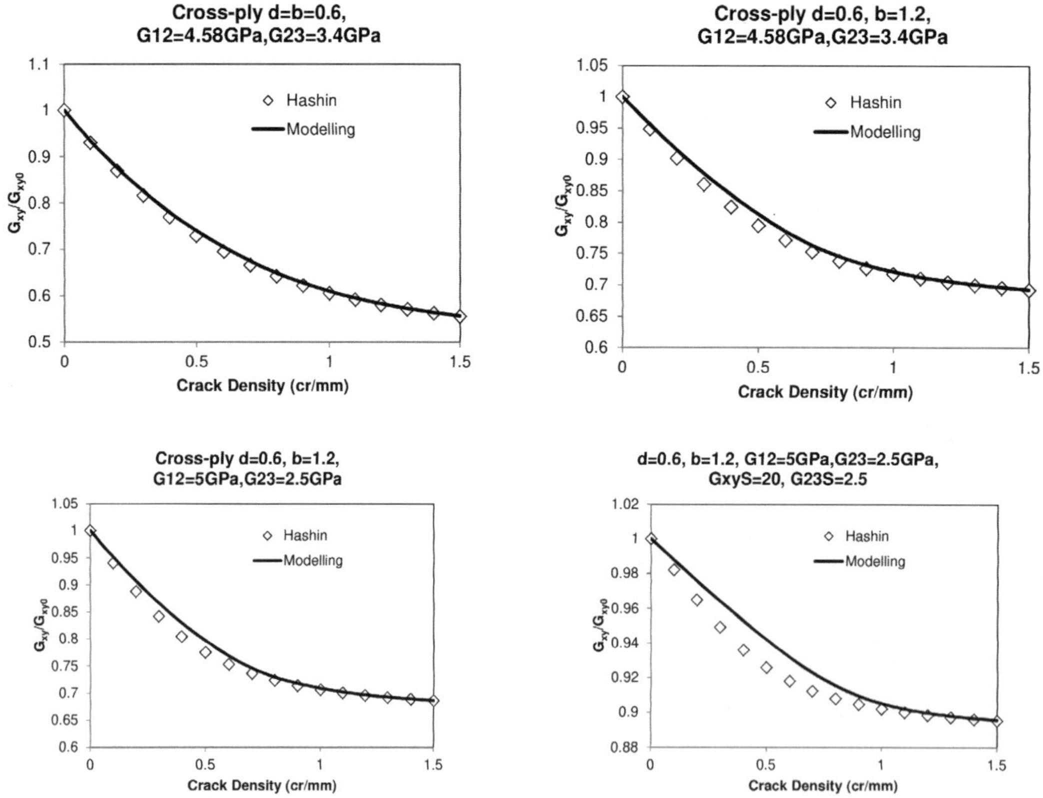

Figure 3: Shear Modulus of damaged: a-c: Cross-ply laminate, d: laminate with a sublaminate on the surface layers

3.2 Parametric analysis

First, the thickness of the outer 0° layer was kept constant and equal to 0.3mm (similar to the thickness of a typical non-aerospace grade carbon fibre prepreg) and the thickness of the 90° was given different values. Specifically, values from 0.01-3mm were studied (the very small unrealistic values were necessary to use to reveal the trends). The step of the increase of the 90-layer thickness was 0.3mm, ie the same as prepreg tape thickness. For this analysis, the values of the in-plane and out-of-plane shear modulus were kept constant as well; G_{12}=4.58GPa and G_{23}=3.4GPa. The results are presented in figure 4. The arrow in figure 4 indicates the direction of the thickness increase; larger thickness leads to higher G_{xy} reduction. It is shown that the thickness of the transverse layers affects both the saturated value of G_{xy} for high crack densities and the slope of the curves. One more conclusion is that the saturation in the crack density comes earlier as the thickness increases. That proves the theory saying that the crack spacing cannot be smaller than the thickness of the 90° plies.

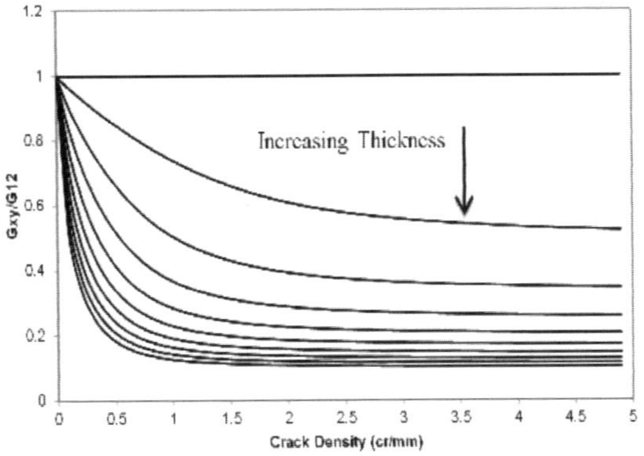

Figure 4: Effect of 90° thickness on the shear modulus of a cross-ply laminate. The thickness of the 90-layer $d \in [0.01,3]mm$**, 0-layer thickness 0.3mm**

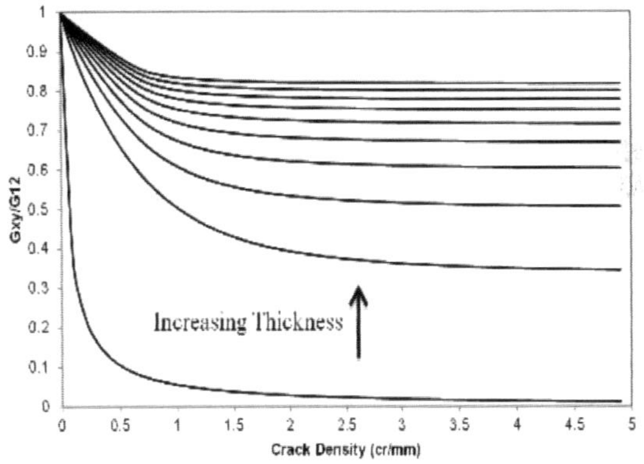

Figure 5: Effect of 0° thickness on the shear modulus of a cross-ply laminate. The thickness of the 0-layer $b \in [0.01,3]mm$**, 90-layer thickness 0.3mm**

The same methodology was followed in order to quantify the effect of the thickness of the 0° plies. The thickness of the transverse layer was kept constant and equal to 0.3mm and thickness of the longitudinal layer was given values from 0.01 to 3mm with an increment of 0.3mm. The results are presented in figure 5.

Figure 5 shows the significance of the longitudinal plies on the final shear properties of the cross-ply. That becomes obvious since for 3mm thick 0° plies, the shear modulus reduction due to cracking is almost 10% at saturation, whereas for 0.3mm (ie only one 0° ply) the reduction is close to 53%.

Next, the results for GF/EP laminates were compared to results in [11] from Finite Element analysis. The parameters used for the modeling are given on table 1. The analysis involved three different stacking sequences; $[0_n/90_2]_s$, where n=1, 2 and 4.

Table 1: Parameters used for the comparison

Material	G12 (GPa)	G23 (GPa)	d (mm)	b (mm)
GF/EP	5.0	5.36	0.1	0.1

The results presented in figure 6, obtained using the two different methods are still rather different. As expected the current model gives lower values than FEM. We may conclude that performed refined stress analysis is still not sufficient and should be improved by considering z-dependent in-plane shear stresses also in the cracked layer.

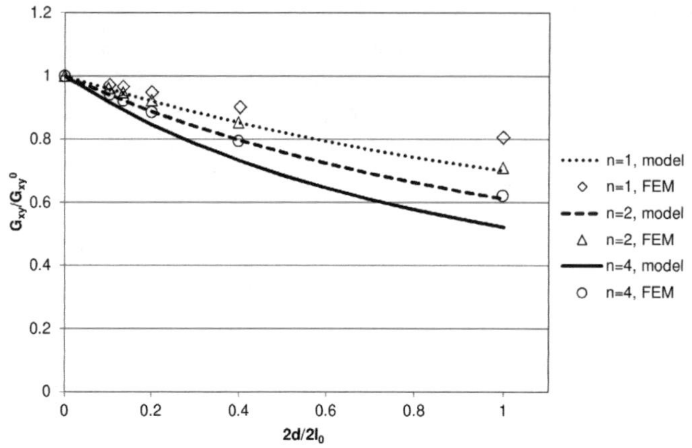

Figure 6: Comparison between Finite Elements results and modeling for $[0_n/90_2]$ GF/EP laminate

4. Conclusions

In the current study, analytical model for shear modulus reduction of laminates due to intralaminar cracking in 90-layer was developed. In contrast to the respective model suggested by Hashin, the dependence of the stress distributions on the thickness coordinate is described by shape function with arbitrary parameter defined in the process of the minimization of the complementary energy. It is proven that the current model gives more accurate predictions than the Hashin's model but the improvement is modest. The performed parametric analysis shows the significance of the individual ply thickness and the out-of-plane shear modulus on shear modulus reduction. Finally, the output of the model is compared with finite element results concluding that further improvement of the variational model by more refined stress analysis in the cracked layer is required.

5. References

[1] Hashin Z, 1985, Analysis of cracked laminates: a variational approach, *Mech of Mater* **4** 121-36

[2] Tan SC, Nuismer RJ, 1989, A theory for progressive matrix cracking in composite laminates, *J Comp Mater* **23** 1029-47

[3] Abdelrahman WG, Nayfeh AH, 1999, Stress Transfer and stiffness reduction in orthogonally cracked laminates, *Mech of Mater* **31** 303-16

[4] Jinghong F, Junqian Z, 1993, In-situ damage evolution and micro/macro transition for laminated composites, *Comp Sci Technol* **47** 107-18

[5] Tsai CL, Daniel IM, 1992, The behaviour of cracked cross-ply composite laminates under shear loading, *Int J Sol Struct* ***29*** **(24)** 3251-67

[6] Zhang J, Fan J, Soutis C, 1992, Analysis of multiple matrix cracking in [±θm/90n]s composite laminates. Part 1. In-plane stiffness properties, *Composites* **23(5)** 291-304

[7] Tsai CL, Daniel IM, Lee JW, 1990, Progressive matrix cracking of crossply composite laminates under biaxial loading. Microcracking-induced damage in composites, *Ed G.J. Dvorak and D.C. Lagoudas, Proc. Of ASME: 1990 Winter Annual Meeting. AMD-vol.111* (**22**) 9-18

[8] Kashtalyan M, Soutis C, 2000, Stiffness degradation in cross-ply laminates damaged by transverse cracking and splitting, *Composites Part A* **31** 335-51.

[9] Gudmundson P, Zang W, 1993, A universal model for thermoelastic properties of macro cracked composite laminates, *Int J Sol Struct* **30** 3211-31.

[10] Lundmark P, Varna J, 2003, Modeling thermo-mechanical properties of damaged laminates, *Key Eng Mater* **251-252** 381-387.

[11] Lundmark P, Varna J, 2006, Crack face sliding effect on stiffness of laminates with ply cracks, *Composites Science and Technology* **66** 1444–54.

Characterization of damaged composite laminates by an optical measurement of the displacement field

M S Loukil[1, 2], J Varna[2] and Z Ayadi[1]

[1] Institut Jean Lamour, (SI2M), EEIGM 6 Rue Bastien Lepage, F-54010 Nancy, France
[2] Division of Polymer Engineering, Luleå University of Technology, SE-97187 Luleå, Sweden

E-mail: mohamed-sahbi.loukil@eeigm.inpl-nancy.fr

Abstract. The degradation of the elastic properties of composite laminates with intralaminar cracks is caused by reduced stress in the damaged layer which is mainly due to two parameters: the crack opening displacement (COD) and the crack sliding displacement (CSD). In this paper these parameters are measured experimentally providing laminate stiffness reduction models with valuable information for validation of used assumptions and for defining limits of their application. In particular, the displacement field on the edges of a $[0/ +70_4/ -70_4]_s$ glass fiber/epoxy laminate specimens with multiple intralaminar cracks is studied and the COD and CSD dependence on the applied mechanical load is measured. The specimen full-field displacement measurement is carried out using ESPI (Electronic Speckle Pattern Interferometry). By studying the displacement discontinuities, the crack face displacements were measured. A comparison between the COD and the CSD (for the same crack) is performed.

1. Introduction

Composite materials have found usage in many industrial applications and more recently they are increasingly being used in aerospace panels and airframes. The use of composites in the aerospace industry is justified by their excellent specific modulus and strength (referred to the property divided by the density).

When a composite is loaded in tension with increasing load it will eventually fail (macroscopically). The failure is preceded by initiation and evolution of several microdamage modes. On the microscale, a part of the matrix can fail, fibers can fail and there can be fiber/matrix interface debonding (Figure 1a). On the mesoscale, the first mode of damage is usually intralaminar cracking in off-axes layers with respect to the main load component (Figure 1b). These cracks run parallel to fibers in the layer with the crack plane being transverse to the laminate midplane. Intralaminar cracks do not usually cause the final failure of a laminate, but may significantly impair the effective properties of the composite and serve as a source for other damage modes initiation, such as delamination and microcracking in the adjacent plies (Figure 1c). Due to this kind of microdamage (transverse crack) a composite undergoes stiffness reduction when loaded in tension.

Many theoretical analytical and numerical models have been developed to calculate the reduced stiffness [1-7]. The opening and sliding of crack surfaces reduce the average strain and stress in the damaged layer, thus reducing the portion of the load carried by this layer. It results in reduction of the laminate thermo-elastic properties. In several papers [1,3-7] the COD and CSD have been analyzed in the context of stiffness reduction. All of them are based on idealized assumptions, for example, assuming linear elastic material behavior even in high stress concentration region at crack tip as well as linearity in shear loading and assuming idealized geometry of these cracks which would not change during the service life [2].

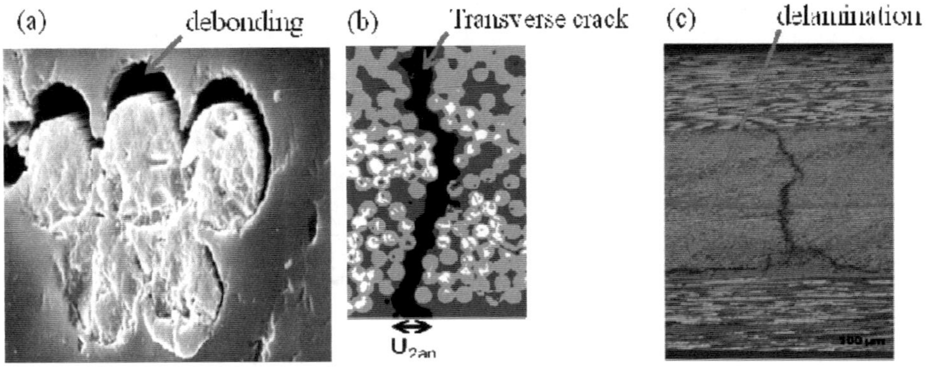

Figure 1. Example of damage modes in composites.

In a linear model these quantities are proportional to the applied load and, therefore, the COD and CSD should be normalized to be used in stiffness modeling. The effect of material properties on the normalized COD was studied experimentally using optical microscopy of loaded damaged specimens in [3,4]. It was shown that the measured COD profiles and average values are affected by the constraining layer orientation and stiffness.

To establish the link between the damaged laminate thermo-elastic constants and the microdamage parameters (COD and CSD) a theoretical framework studies was developed in [1,5]. It was shown that the details of the relative displacement profile along the crack surface are not important: only the average values of these quantities enter the stiffness expressions directly.

The effect of geometrical parameters and properties of the surrounding layers on the COD and CSD was studied using FEM in [1,5]. Analysis revealed that only two parameters are of importance for the normalized average COD and CSD: the modulus ratio of the cracked and the support layers in direction transverse to the crack and the ratio of their thickness. Based on this analysis simple empirical relationships (power laws) were suggested.

The COD and CSD dependence on crack density was analyzed theoretically in [6] and studied using FEM in [7]. It was shown that if the crack density is high the stress perturbations of two neighboring cracks interact and the average stress between cracks at the given applied load is lower. It means that the COD and CSD of interacting cracks are smaller than for non-interactive cracks.

All these studies and analyses assume a linear elastic material with idealized geometry of cracks. The only correct way to validate these assumptions is through experiments. To study experimentally the COD and the CSD, the full-field measurement technique (ESPI) [8,9] was used in this work.

The main objectives of the present study are to visualize the displacement field on the edge of a $[0, +70_4, -70_4]_s$ Glass Fiber/Epoxy (GF/EP) laminate specimen with multiple transverse cracks and to analyze the crack opening displacement dependence on the applied mechanical load. Using the displacement map, it is possible to obtain the displacement profiles along the tensile-axis. The different profiles drawn along the specimen edge at several distances from the mid-plane correspond to the different plies. By studying the displacement discontinuities, we can measure the crack opening displacements corresponding to the cracks in the measurement field. A comparison between the COD and the CSD (for the same crack) is given in this paper.

2. Experimental technique and material

2.1. Experimental technique: ESPI

The principle of ESPI is based on the interference of two coherent laser beams: reference beam and observation beam. For in-plane (x-z plane) measurement, the two beams are reflected and scattered by the rough measuring surface; for out-of-plane measurement (y-direction), only observation beam is reflected by the surface while the reference beam goes to the sensor directly.

The two beams are overlaid and produce a speckle pattern, which is recorded by a Charge Coupled Device (CCD) camera. Figure 2 shows a schematic setup for the in-plane deformation measurement by ESPI.

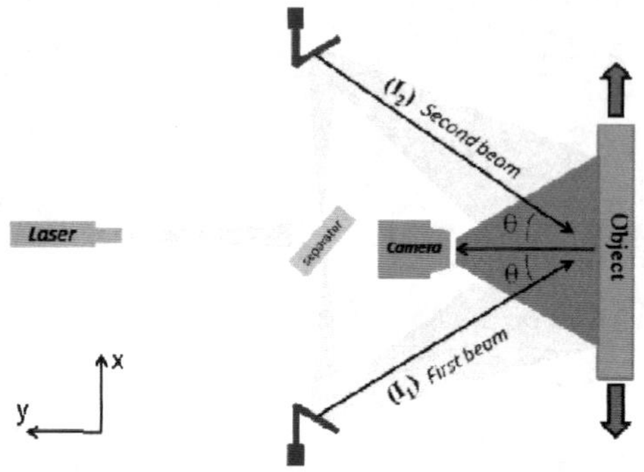

Figure 2. Schematic of ESPI setup.

The intensity, I_m of the speckle interference pattern is

$$I_m = I_1 + I_2 + 2\sqrt{I_1 I_2}\, cos(\phi + \varphi_m) \tag{1}$$

where I_1 and I_2 are the intensities of the two coherent beams, ϕ is the interference phase of the two coherent beams, and φ_m is the introduced phase-shift.

By subtracting two speckle patterns at different states, before and after deformation, interference fringes can be obtained. Using the four-step phase shifting method [10,11], the interference phase at each state can be determined by four intensities with phases separated by $\pi/2$. Assuming the interference phases before and after the deformation are ϕ and ϕ' respectively, the phase difference $\Delta\phi$ can be obtained by $\phi - \phi'$. Applying the phase-shift process, the displacement map can be obtained.

The in-plane displacement U_x (or U_z) can be calculated by

$$U_x = \frac{\lambda \times \Delta\phi}{4\pi \sin\theta} \tag{2}$$

where θ is the incidence angle of the illumination beams, λ is the wavelength of the applied laser, and $\Delta\phi$ is the phase difference between the two states: before deformation and after deformation.

Similarly, the out-of-plane displacement U_y can be obtained by

$$U_y = \frac{\lambda \times \Delta\phi}{2\pi(1 + cos\,\theta)} \tag{3}$$

ESPI can provide full-field deformation information of different materials with a very high-precision. However it is a very sensitive technique and the results can be easily influenced by external factors such as noise and vibration. For that reason tests should be carried out in relatively quiet circumstances.

2.2. Material

The [0, 70_4,-70_4]s laminate (see the geometrical details in Figure 3) was made of glass fiber/epoxy using vacuum bag technique. The thickness of the laminate is 2.8 mm. The elastic properties obtained in tests on unidirectional and angle-ply composite specimens are:

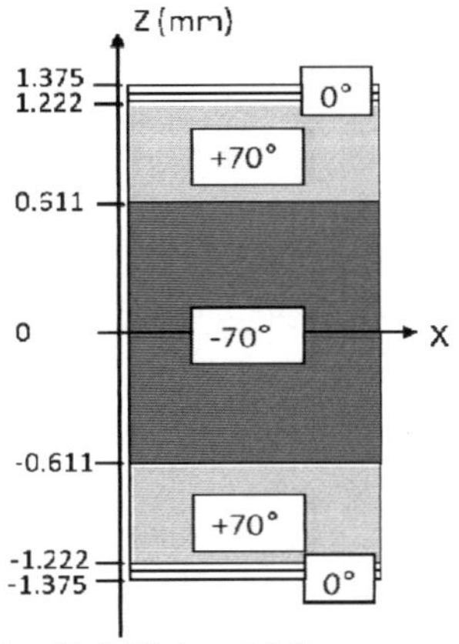

Figure 3. Shematic view of the $[0,70_4,-70_4]$s laminate lay-up. (Z-axis is in the thickness direction).

Longitudinal modulus E_1(GPa) = 44.7
Transverse modulus E_2(GPa) = 12.7
Poisson's ratio $\nu_{12} = 0.297$
In-plane shear modulus G_{12}(GPa) = 5.8
Specimens of 20 mm width were reinforced with GF/EP end tabs in the gripping area.
A free length between tabs was 110 mm. The tensile loading to introduce damage in form of microcracks was applied using an INSTRON tensile testing machine with displacement rate of 3 mm/min. Cracks initiate first in the -70 layer, which is two times thicker than the +70 layer and propagate through the whole width of the specimen. At higher applied stresses, cracks appear in the +70 layer. The presence of these cracks (in +70 layer) affect the opening displacement of the large cracks in the -70 layer.

3. Results

3.1. Normalized relative displacement measurement
The specimen edge is illuminated by two beams which come from the same laser that has $\lambda = 0.638\mu m$ as wavelength. These beams illuminate the object at equal but opposite angles $\pm\theta$ to the surface normal (see figure4). Using this method of illumination, the measurement direction is in the x-axis direction which is the tensile-axis.

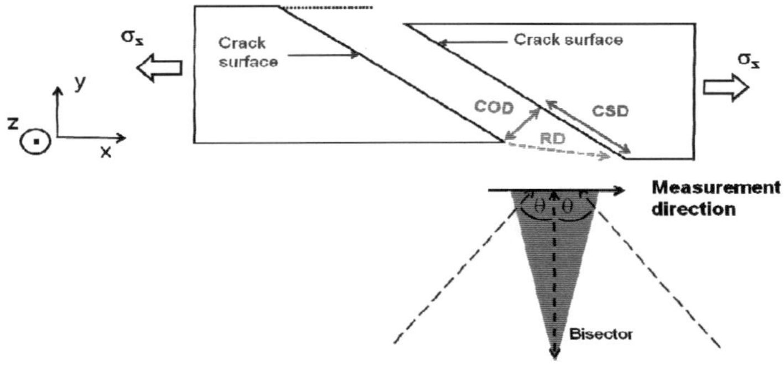

Figure 4. Symmetric illumination.

The digitization of the $10mm \times 10mm$ image is performed at 512×512 pixels, thus the pixel size is $d = 19.5\mu m$. To get the displacement map, the image corresponding to the phase difference was averaged using a $[11 \times 11]$ filter kernel.

Figure 5 shows the X-displacement profile in the central ply (-70°) when the cracks were first initiated. The displacement discontinuities caused by the cracks are clearly visible.

These displacement jumps indicate the presence of two cracks in the region of study. The displacement slope (strain) is very small in the area corresponding to the cracks surfaces where according to the boundary condition the crack surfaces are stress free.

Because of the 70° fiber orientation, the crack displacement discontinuity measured on the edge is not directly COD and neither the CSD. It is a projection of both these displacements on the specimen edge plane (Figure 4). By definition, the crack opening displacement (COD) is transverse to the fiber direction and the sliding displacement (CSD) is in the fiber direction. The displacement jump is defined as a relative displacement (RD) which is related to the COD and the CSD by this relationship:

$$RD_x = COD \times \sin\theta + CSD \times \cos\theta \quad (4)$$

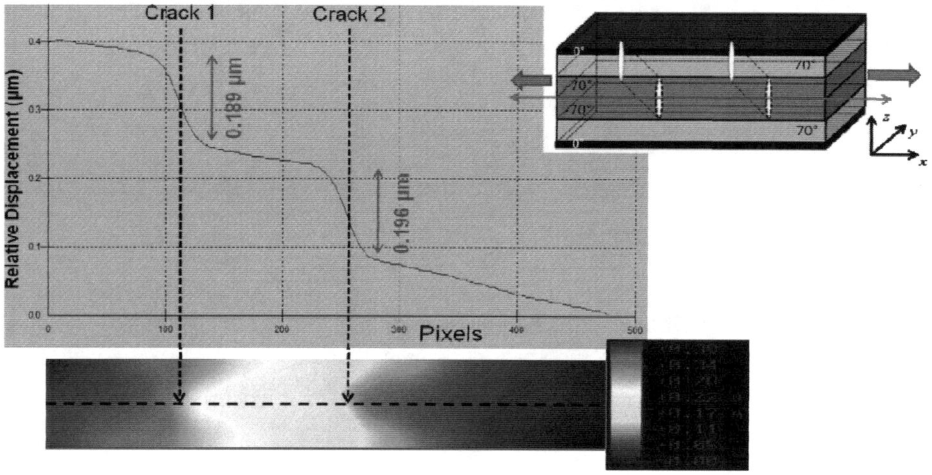

Figure 5. Profile of the X-direction relative displacement along the mid-plane (on the specimen edge) corresponding to a variation of the relative average stress $\Delta\sigma = 0.497 MPa$.

The displacement gap observed on the edge corresponds to a certain change in the applied average stress $\Delta\sigma$. In other words the presented displacement data correspond to this load change.

In order to perform comparative analysis for several applied load levels the relative displacement (RD) has to be normalized with respect to the relative average stress $\Delta\sigma$. Thus, the results presented in the following for the normalized relative displacement (RD_n) correspond to 1 MPa of the applied stress to the laminate, where the upper index n is used to indicate the normalized values.

For an average stress increase from 27.01MPa to 27.51MPa a $0.189\mu m$ and $0.196\mu m$ relative crack face displacement was measured for the first and the second crack, respectively. It corresponds to $RD_{1n} = 0.380\mu m/MPa$ and $RD_{2n} = 0.394\mu m/MPa$ for the normalized crack opening displacement at the averaged applied stress $\Delta\sigma = \frac{27.01 + 27.51}{2} = 27.26 MPa$.

Using the displacement map, the displacement profiles were drawn (not shown) along the tensile-axis at two different distances from the mid-plane (in the middle of +70° and 0° layers). Two cracks were detected on the edge of the +70° layer very close the cracks in the-70° layer. The normalized relative displacement of these cracks which run along fibers in the +70° layer was smaller than those in -70° layer and interface delamination between both crack systems was observed.

For 0° layer, the displacement profile does not show any displacement jumps, which means there are no cracks in this layer at this applied stress level.

The separation of the values of the COD and the CSD can be performed using the following two methods:

- To measure the relative displacement RD with θ' as new incidence angle of illumination beams $(\theta \neq \theta')$. Hence, using equation (4) the RD can be calculated with two different angles θ and θ' and then the COD and the CSD can be deduced.
- To measure directly the COD. This method is used and explained in the following section.

3.2. Normalized crack opening displacement

The object is illuminated by two beams. To measure directly the COD, theses two beams have to illuminate the object at equal and opposite angles to the fiber direction axis, which means the bisector of the angle between the first and the second beam has to be in the same direction as fibers of the studied layer (Figure 6). In this case, the measurement direction is transverse to the fiber direction. Hence, using this method of illumination, we can measure directly the COD.

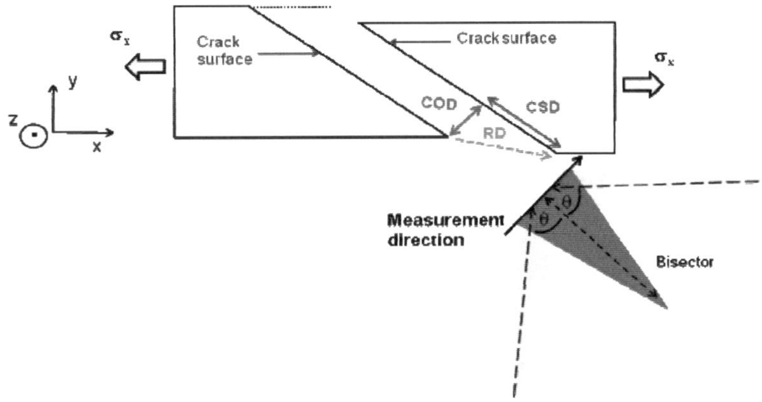

Figure 6. Non symmetric illumination.

For the same crack, we measured the normalized relative displacement RD_n and the normalized crack opening displacement COD_n separately, and then we deduced the normalized crack sliding displacement CSD_n using equation (4).

All results are presented in this following table.

Table1. A comparison between the COD and the CSD.

	RD_n (μm/MPa)	COD_n (μm/MPa)	CSD_n (μm/MPa)	**COD/CSD**
Crack 1	0.380	0.300	0.286	**1.049**
Crack 2	0.394	0.311	0.297	**1.047**

For the analyzed GF/EP laminate, Table 1 shows that the COD is slightly bigger than the CSD.

4. Conclusion

The displacement field on the edge of a $[0/+70_4/-70_4]_s$ glass fiber/epoxy laminate specimens with two intralaminar cracks is studied and the crack face opening displacement (COD) and crack face sliding displacement (CSD) change with the applied mechanical load is measured.

The presented results prove that ESPI can be used to measure directly the crack opening displacement and deduce the crack sliding displacement. A comparison between these two parameters shows that the COD is slightly bigger than the CSD.

The COD and the CSD depend on the materials properties and layer orientation and thickness which means that the ratio (COD/CSD) varies. As future work, we are planning to study more laminates with different lay-ups and material properties.

The measurement of COD and CSD at low and high crack density is planned to investigate the effect of crack density.

5. References

[1] Lundmark P and Varna J 2005 Constitutive relationships for laminates with ply cracks in in-plane loading *Int. J. Damage Mech.* **14 (3)** 235-61
[2] McCartney LN and Schoeppner GA 2000 Predicting the Effect of Non-uniform Ply Cracking on the Thermo-elastic Properties of Cross-ply Laminates *Comp. Sc. and Tech.* **62** 1841-56
[3] Varna J, Berglund LA, Talreja R and Jakovics A 1993 A study of the crack opening displacement of transverse cracks in cross ply laminates *Int. J. Damage Mech* **2** 272–89
[4] Varna J, Joffe R, Akshantala NV and Talreja R 1999 Damage in composite laminates with off-axis plies *Comp. Sc. Tech.* **59** 2139–47
[5] Lundmark P and Varna J 2006 Crack face sliding effect on stiffness of laminates with ply cracks. *Comp. Sc. Tech.* **66** 1444–54
[6] Joffe R, Krasnikovs A and Varna J 2001 COD-based simulation of transverse cracking and stiffness reduction in [S/90n]s laminates. *Comp. Sc. Tech.* **61** 637–56
[7] Lundmark P and Varna J 2011 Stiffness reduction in laminates at high intralaminar crack density: effect of crack interaction *Int. J. Damage Mech.* **20** 279-97
[8] Farge L, Ayadi Z and Varna J 2008 Optically measured full-field displacements on the edge of a cracked composite laminate *Comp. Part A* **39** 1245-52
[9] Farge L, Varna J and Ayadi Z 2010 Damage characterization of a cross-ply carbon fiber/epoxy laminate by an optical measurement of the displacement field *Comp. Sc. and Tech.* **70** 94-101
[10] Moore AJ and Tyrer JR 1996 Two-dimensional strain measurement with ESPI. *Opt. Laser Eng.* **24** 381-402
[11] Nakadate S and Saito H 1985 Fringe scanning speckle interferometry *Appl. Opt.***24(14)** 2172-80

IOP Conf. Series: Materials Science and Engineering **31** (2012) 012023 doi:10.1088/1757-899X/31/1/012023

The effect of mesostructure heterogeneity on cracks initiation and the displacement distribution in NCF composites

H ZRIDA [1,2], K GIANNADIS[2], J VARNA [2] and Z AYADI [1]

[1] Institut Jean Lamour, Nancy Université, Ecole Européenne d'Ingénieurs en Génie des Matériaux, 6 rue Bastien Lepage, F-54010 Nancy, France
[2] Division of polymer Engineering, Lulea University of Technology SE-97187 Luleå, Sweden

E-mail : Hana.zrida@eeigm.inpl-nancy.fr

Abstract. Non Crimp Fabrics (NCF) are promising new generation composite materials. They are now being used in some sections of composite industry, for example in wind turbine blades and boat hulls. The aerospace industry also shows an increasing interest in this material, thanks to the low cost of its manufacturing process.
NCFs are special types of textile composites, made of layers of parallel fiber bundles oriented in different directions and separated by resin.
Due to the manufacturing process the fiber bundles are not perfectly straight. They show a certain degree of waviness which decreases the stiffness and the strength of the material. The heterogeneous mesostructure affects the mechanical properties of the material and the failure mechanisms. This was studied using both numerical and experimental methods.
In our experimental approach, a carbon fiber/ epoxy resin laminate with uniform fiber distribution was manufactured by voluntarily introducing waviness to simulate the NCF composites. The displacement map was studied against the thickness of a sample loaded in tension, using ESPI (Electronic Speckle Pattern Interferometry),
This can give us a primary idea of the micro damage initiation and the cracks' shapes.

1. Introduction

The main interest of the aeronautic industry is not only to reach the minimum weight by keeping a good performance, but also and most importantly, to achieve at lowest cost of the best mechanical properties.
The new composites NCFs (Non Crimp Fabrics), with fiber bundle structure, meet the industry's expectations by providing a significant savings in terms of costs when compared to the cost of the usual composites with uniform fiber distribution.
NCFs are a relatively new class of textiles in which a fabric is constructed of layers of tow aligned in specific directions [1-2]. This type of textiles differs from the usual fabrics by the use of a secondary polyester yarn that binds the fiber bundles together according to straight directions (see figure 1). The dry preform is impregnated with resin infusion. This manufacturing process is realized by Techtextiles International Ltd employing Liba technology [3].

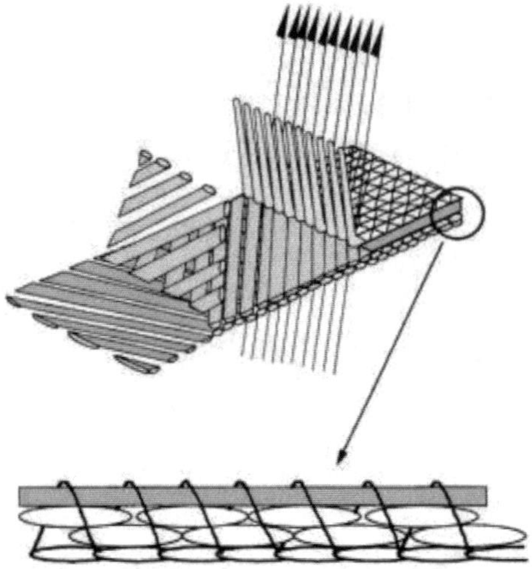

Figure1. Principle of NCF's manufacturing process.

The economical factors were discussed in [3]. The deposition rate of the layers for NCF is seven times faster than that for prepregs (pre-impregnated composites), since one NCF's layer with fiber bundle structure is the equivalent of 7 prepreg's layers with uniform fiber distribution. In addition, the step of layer pressing is not necessary in NCF's manufacturing. Thus, NCF production is considerably cheaper than the prepreg production.

Unfortunately, due to this manufacturing process the fiber bundles are not perfectly straight, showing a certain degree of waviness which affects the mechanical properties and the damage mechanisms [4-5]. The goal is therefore to find a balance between the cost and the performance.
The main objective of the present study is to determine the effect of waviness on micro damage initiation in NCF composites through an experimental approach using Speckle interferometry during tensile test.

2. Experimental approach

Many parameters can affect the damage mechanisms in NCFs, such as the fiber volume fraction inside the bundle, which changes from the centre to the edge of the bundle. Adding to that, the damage mechanisms can be influenced by the presence of the secondary knitting yarn which does not exist either in prepregs or in textile composites. Moreover, new kinds of cracks appear, due to the fiber bundle structure, such as longitudinal cracks [4].

Our interest is to study the effect of the waviness on the mechanical properties of NCFs and in the present work we tried to do it in a controlled way. That is why we have simplified the present experimentation by preparing a laminate with a band of composite prepreg made of carbon fiber and epoxy resin. Waviness was introduced to simulate the mesostructure imperfections that occur in NCFs because of the knitting yarn. Tensile test and Speckle interferometer observations were conducted to study the form of cracks and the displacement variations related to the waviness.

2.1. Manufacturing process and materials

The manufacturing process of the cross ply laminate was performed according to the following steps: First, three layers of carbon fiber/ epoxy resin prepreg were put in [0,90,0°]. Then, the whole material was covered with a release film to avoid any resin spread. Next, straight and equidistant wires were fixed and put on both sides of the laminate. Finally, the assembly was covered, put under vacuum and pressed with a hot pressure machine at T=120°C under P = 3bar for one hour. The plate obtained, holding the wires' mark assumed as waviness, was cut into rectangular samples perpendicularly to the

IOP Conf. Series: Materials Science and Engineering **31** (2012) 012023 doi:10.1088/1757-899X/31/1/012023

wires' direction (see figure2). Two kinds of wires were used in order to get two different waves with different amplitudes and wavelengths λ.

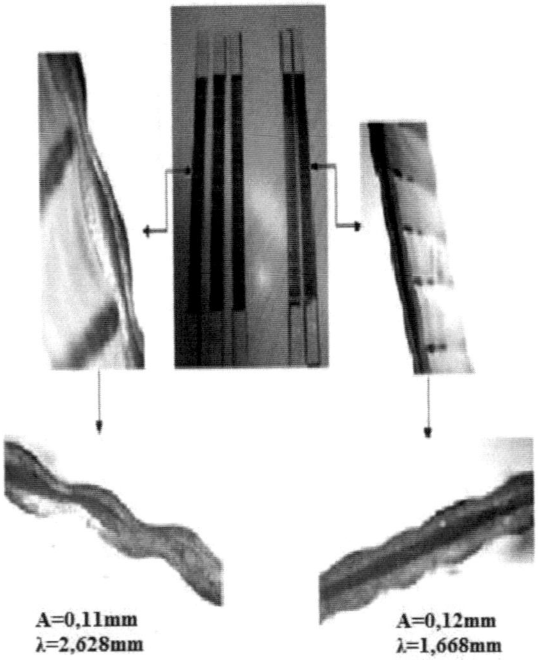

Figure2. Samples containing waviness with different amplitudes A and wavelengths λ.

The elastic properties of the unidirectional composite are as follows:
Longitudinal modulus: E_1 (GPa) = 118
Transverse modulus: E_2 (GPa) = 9
Poisson's ratio: $\nu_{12} = 0.3$
In plane shear modulus: G_{12} (GPa) = 4
Prepreg tape thickness: 0.33mm
Specimens with a dimension of 120mm x 10mm were cut and their edges were polished. The thickness varies between 0.6 and 1mm.

2.2. Tensile test, stiffness degradation and crack's observation

A specimen containing waviness with an amplitude of A=3.2mm and wavelength λ=0.1mm was loaded in tension until failure (see figure3) and the longitudinal displacement (Ux) was measured by an extensometer.

IOP Conf. Series: Materials Science and Engineering **31** (2012) 012023 doi:10.1088/1757-899X/31/1/012023

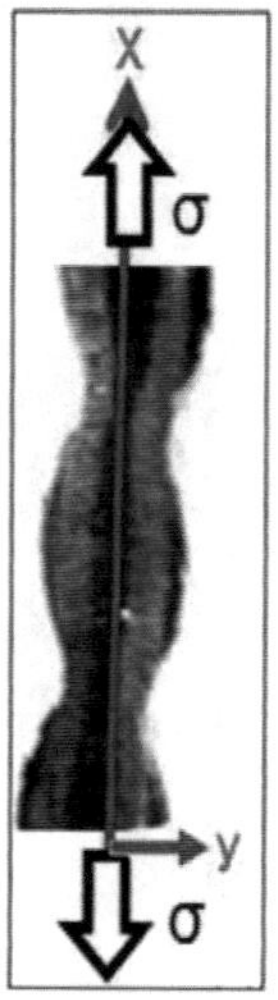

Figure3. A [0,90,0] waved laminate loaded in tension.

The stress- strain curves were depicted for the waved specimen and for a reference which has the same geometrical configuration but without waviness. According to figure 4, the stiffness is smaller in the sample (with waviness) (E_x=64,5 GPa) than in the reference (without waviness) (E_x=71,7 GPa). This stiffness degradation (along x) in the waved sample is related to the decrease of E_x in the 0° layers. In fact, in the reference E_x of the 0° layer corresponds to the longitudinal modulus of the material (E_1) and E_y corresponds to the transversal modulus E_2. In the 90° layer E_x corresponds to E_2 and E_y corresponds to E_3. However, in case of waviness in (x,y) plane, in 0° layer, E_x becomes a combination of E_1 and E_2, so it is smaller than for a flat layer. In contrast, in a 90° layer, E_x doesn't change because it is a combination of E_2 and E_3 which have the same value. Consequently, the stiffness degradation is mainly due to the decrease of E_x in the 0° layer.

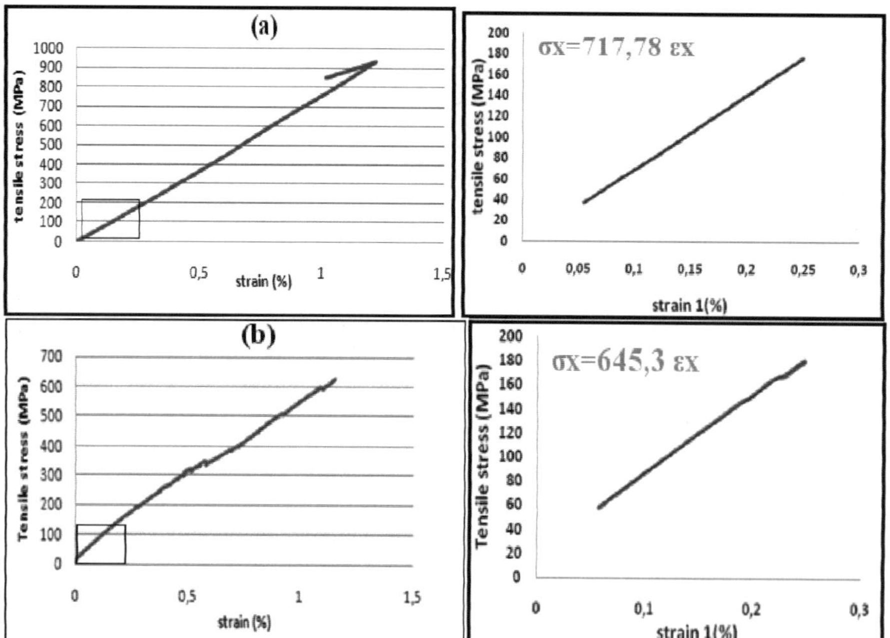

Figure4. Stress-strain behavior for: (a) a reference (without waviness) (b) a specimen (with waviness).

Another tensile test was carried out. Cracks were introduced and then observed by an optical microscopy. Figure 5 shows the cracks that occur and which were observed throughout the thickness of the sample. We have noticed that the cracks seen are not ordinary transverse cracks. In other words, the cracks' direction is not perpendicular to the loading direction like it is used to see in flat cross ply laminates. Moreover, the angle between the crack and the loading direction is different from one wave to another as can be observed in figure6.

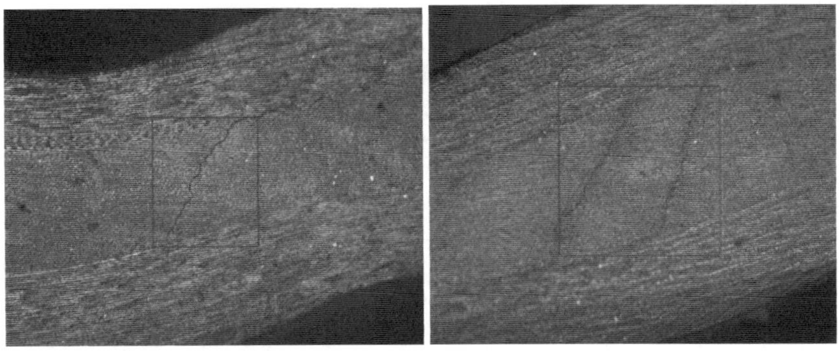

Figure5. Cracks in the 90° layer of the cross ply laminate with waviness

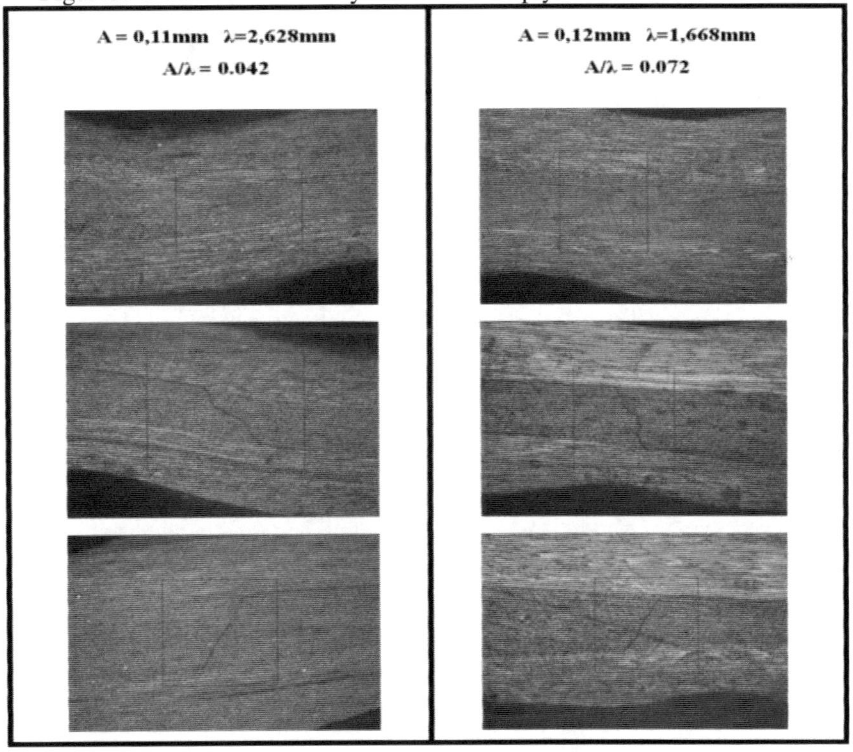

Figure6. Pictures of cracks in 90° layer in two different waved specimens.

In order to understand this special shape of the cracks, we performed a full field displacement measurement using Electronic Speckle Pattern Interferometer (ESPI) and finite element calculations.

2.3. ESPI and full field displacement measurement

ESPI is a full field displacement measurement method which enables us to measure the displacement components in different directions (in plane and out of plane). The displacement measurement using ESPI is around 1-2 µm, which corresponds to the scale of the crack opening displacement size. This is why this technique is used to characterize the damage in some materials like composites and steels…
The ESPI is based on the speckle interferometry. The speckle is the random distribution of the laser light after being reflected by a rough surface [6] (see figure7).

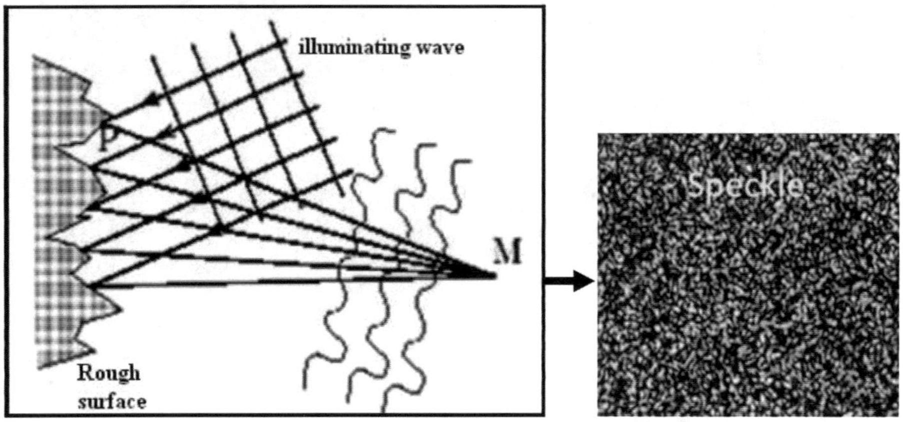

Figure7. Speckle obtained by reflected light

The basic principles of the ESPI technique are described in detail by Farge et al.[Ref7-8].
In the present study, ESPI was used to determine the displacement variation due to the waviness. The experimental arrangement is represented in figure 8. The sample was illuminated by two symmetric laser beams, the corresponding speckles interfere. The interference figures are similar to the speckle figures (see figure9). The speckle of the intensity obtained before loading was subtracted from the one obtained after loading. The phase shift map obtained was averaged using a [11 x 11] kernel filter and then demodulated. Finally, the displacement map was obtained and the x displacement component (U_x) was measured in different positions in the sample thickness. We determined the U_x map for a loading between 68,32 MPa and 71,16 MPa, in other words, for $\Delta\sigma = 2{,}84$ MPa.

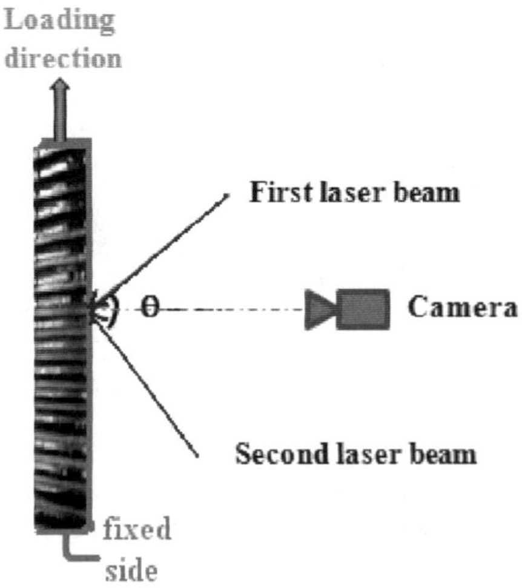

Figure8.The experimental arrangement for displacement determination using ESPI

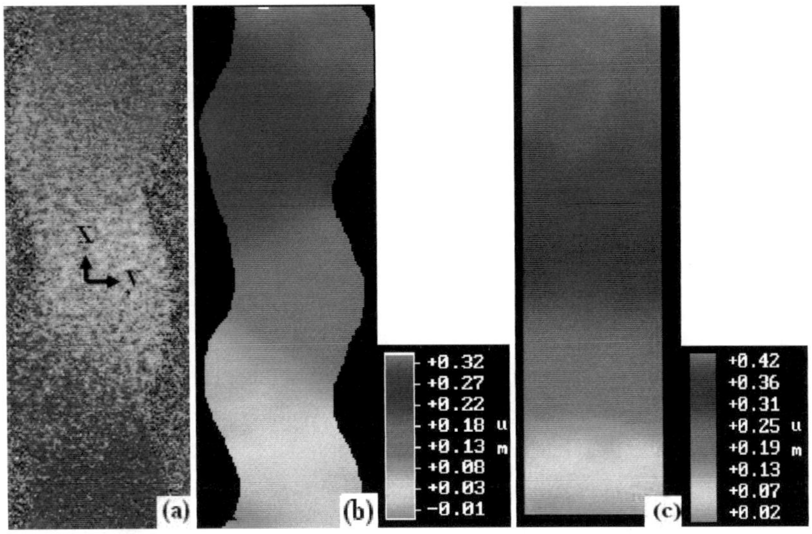

Figure 9.(a) Fringes map after subtraction (b) Displacement map for a laminate with waviness (c) Displacement map for a laminate without waviness

Before doing any measurement, a simple observation shows that the fringes, which characterize the same x displacement, are not horizontal and parallel in the waved laminate. This observation differs from what used to be seen in flat laminates. When there is no waviness, all points placed at fixed position along x have the same displacement along y. However, with the existence of the waviness, U_x behaves differently, as we can see in figure 10 and more clearly with the superposition in figure 11. If we consider a single wavelength, we notice that in the beginning, the middle and the end of the wave, which correspond to the positions marked respectively by the pixels 171, 268 and 385, the curves are horizontal. That means all points have the same displacement U_x. However, when the undulation shows a certain angle in reference to the direction of tension, U_x curves show a certain slope, and it is no longer constant throughout the sample thickness.

In fact, in tension loading, when the waviness of the two 0° layers are in phase, the structure is more deformed by bending. The p219 section rotates clockwise, which explains why the displacement increases in the left side and decreases in the right side. The p326 section rotates counter-clockwise, which is why we observe the opposite effect.

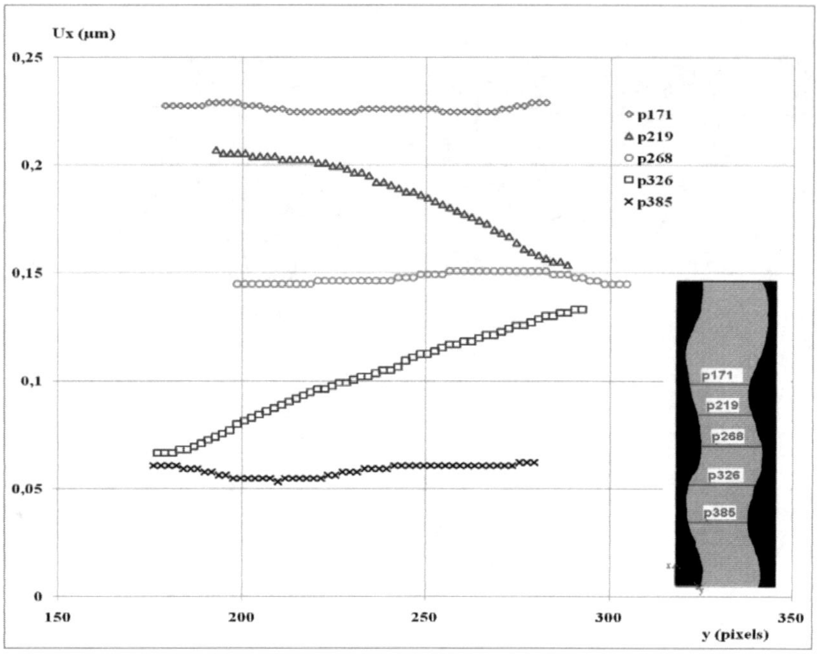

Figure10. U_x behavior along a horizontal section in different positions on the sample thickness

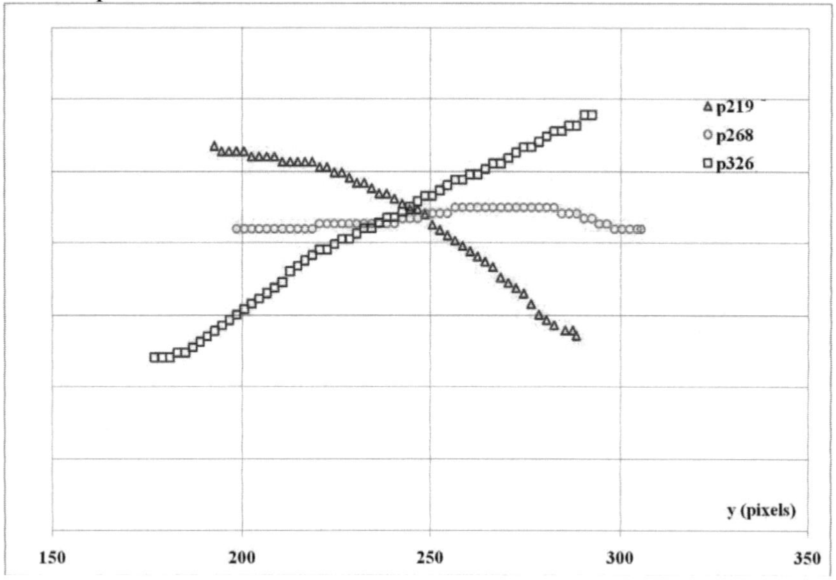

Figure11. Superposition of the previous curves (for p219, p269 and p326)

Consequently, when the composite layers are waved, we see a variation in the displacement profile, which is the effect of the existence of "compressive" and "tensile" sides because of the bending in superposition with overall tension (see figure12).

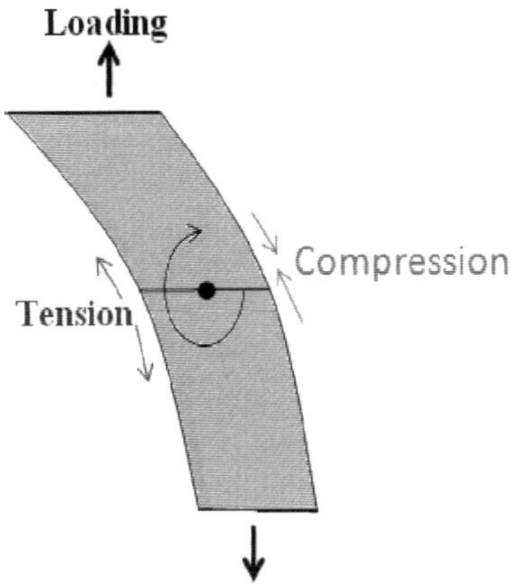

Figure12. Bending deformation for p219 section.

This can prove the shape of the inclined cracks which we have been observed in the previous section after the tensile test. Certainly, shear deformation can appear after the tension loading due to the misalignment of the loading direction with the sample length. To confirm the existence of the shear deformation, the E_{xy} strain should be investigated. For that, the displacement along the y direction U_y has to be measured. The limitation of the speckle is that the U_y measurement is accurate only when the sample thickness is greater than 1 mm. However, the thickness of the sample used in this test is between 0.6 and 1 mm, so the interference fringes cannot be seen.

3. Conclusion

It is shown that the presence of waviness in the layers of the carbon fiber /epoxy composite affects the elastic mechanical properties, a phenomenon which was demonstrated by a simple tensile test. The effect of this mesostructure misalignment was clearly observed on the displacement profile. These displacement variations explain the formation of the inclined cracks in the 90° layer.
The simulation made in this study give a primary idea about what can happen in Non Crimp Fabric composites due to the waviness which occurs during the manufacturing process.

4. References

[1] Hogg PJ, Ahmadnia A, Guild FJ 1993 The mechanical properties of non-crimped fabric-based composites *Comp.***24** 423-32

[2] Hogg PJ, Woolstencrofl DH 1991 Non-crimp thermoplastic composite fabrics: aerospace solutions to automotive problems *7th Ann ASM/ESD Advanced Comp. Conf. Detroit* 339-49

[3] Bibo GA, Hogg PJ Backhouse and Hills A 1998 Carbon-fibre non-crimp fabric laminates for cost-effective damage-tolerant structures *Comp. Sc. And Tech.* **58** 129-43

[4] Edgren F, Mattsson D, Asp LE and Varna J 2004 Formation of damage and its effects on non-crimp fabric reinforced composites loaded in tension *Comp. Sc and Tech.* **64** 675-92

[5] Asp LE, Varna J and Marklund E 2011 Progressive damage Chapter 12 in Non crimp fabric composites : manufacturing, properties and applications *S.Lomov, Woodhead* 289-309

[6] Grédiac M 2004 The use of full-field measurement methods in composite material characterization : interest and limitations *Comp. PartA* **35** 751-61

[7] Farge L Caractérisation de l'endommagement de matériaux composites stratifies à l'aide de la mesure du déplacement par une méthode optique plein-champ,thèse de doctorat,Institut Jean Lamour, Science et Ingénierie des Matériaux et Métallurgie, Energie Mécanique Matériau,(Octobre 2009)

[8] Farge L, Ayadi Z and Varna J 2008 Optically measured full-field displacements on the edge of a cracked composite laminate *Comp. PartA* **39** 1245-52

IOP Conf. Series: Materials Science and Engineering **31** (2012) 012005 doi:10.1088/1757-899X/31/1/012005

Characterization and modeling of performance of Polymer Composites Reinforced with Highly Non-Linear Cellulosic Fibers

L Rozite[1], R Joffe[1,2,3], J Varna[1], B Nyström[2]
[1]Material Science, Luleå University of Technology, S-97187 Luleå, SWEDEN
[2]Materials and Manufacturing, Swerea SICOMP, S-94126 Piteå, SWEDEN

E-mail: liva.rozite@ltu.se, roberts.joffe@ltu.se, janis.varna@ltu.se, byrgitha.nystrom@swerea.se

Abstract. The behaviour of highly non-linear cellulosic fibers and their composite is characterized. Micro-mechanisms occurring in these materials are identified. Mechanical properties of regenerated cellulose fibers and composites are obtained using simple tensile test. Material visco-plastic and visco-elastic properties are analyzed using creep tests. Two bio-based resins are used in this study – Tribest and EpoBioX. The glass and flax fiber composites are used as reference materials to compare with Cordenka fiber laminates.

1. Introduction

There are strong incentives for development of materials from renewable resources. Development of thermosetting composites with different types of natural fibres as reinforcement is one step in the sustainable direction which has been studied by many [1-4]. In recent years a number of bio-based thermosetting resins have become available, making the development of entirely bio-based composites possible [5-7].

It has been shown that natural fibers with high cellulose content (e.g. flax, hemp) compare well in terms of mechanical performance with glass fibers, especially if properties are normalized with density. Advantages of these fibers are low density, renewability, they have environmentally friendly image and have good insulation properties. Therefore, these fibers are considered to be as substitution of glass fibers in automotive, building industries, and probably even possible in marine applications and wind turbine blades.

However, there are certain challenges concerning wide utilization of these fibers in high performance composites, such as poor adhesion between matrix and fibers, limited fiber length, difficulties associated with control of fiber alignment and orientation as well as variability of their properties (depending on location of growth and time of harvest). Therefore, another type of reinforcement of bio-based origin with high cellulose content, continuous Regenerated Cellulose Fibers (RCF), is of high interest. They also can absorb large amount of energy, which is important for toughening composite structures. However, mechanical behaviour of RCF is highly non-linear. As a result, performance of the composite with this reinforcement is also non-linear with the presence of very significant visco-elastic behaviour.

[3] To whom any correspondence should be addressed.

In the present study, new combinations of fibres and resins from renewable resources are used as constituent materials in composites. Fibres made from regenerated cellulose and high quality spun flax fibres are combined with two types of bio-based thermosetting resins.

The long term objective of the work is to predict the mechanical behaviour of these composites by generalizing existing models to capture their time-dependent behaviour. This study is focused on experimental determination of input parameters for these models. In order to identify and quantify parameters needed for the modelling, extensive damage tolerance tests as well as creep experiments are carried out. Preliminary results from tests of composite laminates with RCF show that damage modes in these materials are very similar to those of synthetic composites (like glass and carbon fiber composites). However, these results also show that stress-strain curves for these materials are very non-linear and loading-unloading experiments demonstrate that the shape of the stress-strain curve has the form of a hysteresis loop. Therefore, these materials have to be treated as highly non-linear with presence of visco-elastic and/or visco-plastic effects.

2. Mechanical characterization

2.1. Materials

Two types of fibers were used in this study – regenerated cellulose fibers "Cordenka 700 Super 3" and unbleached high quality flax fibers produced by Linificio e Canapificio Nazionale, Spain, supplied by eXtreme in Italy. Tribest and EpoBioX were used as bio-based resins. Tribest is acrylated, epoxidized soy oil based resin (Cognis, Germany). As a curing agent 2.25% peroxide Benox L40LV (Syrgis Performance Initiators AB) was used. EpoBioX is epoxidised pine oil based resin (Amroy, Finland). As a curing agent for EpoBioX Ca35Tg hardener with mixing ratio 100:27 was used. Tribest resin and EpoBioX base are approximately 75% bio-based.

2.2. Composite manufacturing

Fiber roving was winded on steel plates using a filament winding machine (2 layers for unidirectional composites (UD)). Afterwards fiber preforms were impregnated using vacuum infusion with resin heated to 50°C (Figure 1). Composites with Tribest resin were cured for 8h at 80°C and with EpoBioX resin were cured for 2h at 80°C. Fiber volume fractions of composites are approximately 40-50%.

Figure 1. Filament winding for UD composites (left) and vacuum infusion (right)

2.3. Tensile tests

2.3.1. Composite and neat resin

Composite plates were cut into rectangular shaped specimens and then polished (up to the 1200 grit sand paper). Specimens did not have end tabs but grinding paper (roughness 240) was used to wrap the ends of the specimens in order to prevent specimen crashing in the grips.

IOP Conf. Series: Materials Science and Engineering **31** (2012) 012005 doi:10.1088/1757-899X/31/1/012005

Tensile tests of composites and neat resins were performed on Instron 3366 machine equipped with 10 kN load cell, pneumatic grips and extensometer with a gauge length of 50mm (Instron 2630-111). In all quasi-static tensile tests the speed of the tensile machine's cross-head was 5mm/min which roughly corresponds to a strain rate of 5%/min. Load and displacement was registered by data acquisition unit and stored on PC. Tensile tests were performed at room temperature and relative humidity (RT ~22 °C, RH ~25%). Specimen was clamped in the grips (25 mm from each end) in such a way that gauge length was set to 100 mm.

2.3.2. Single fiber

Single fiber tests were performed according to the ASTM D 3379-75 Standard [8]. Single fibers were manually separated from the bundles. Fiber ends were glued onto a paper frame as shown in Figure 2. Specimens with gauge length of 20 mm were prepared. Tension tests were carried out on an electromechanical tensile machine Instron 4411 equipped with load cell of 5N and pneumatic grips. During mounting, the specimens were handled only by the paper frame. After the clamping of the ends of the paper frame by the grips of the test machine, frame sides were carefully cut in the middle. The tests were displacement-controlled with the loading rate of 2 mm/min (which corresponds to 10 %/min). Images of the fiber in the frame (schematic drawing and photograph) and mounted in the machine with cut paper frame are shown in Figure 2. Strain was determined from the displacement of cross-head of tensile machine and system compliance was taken into account.

Two types of single fiber specimens were tested: fiber with twist and fibers with no twist. Fibers were not specially conditioned, they were stored and tested at room temperature and relative humidity: RT = 23°C, RH ≈ 25%.

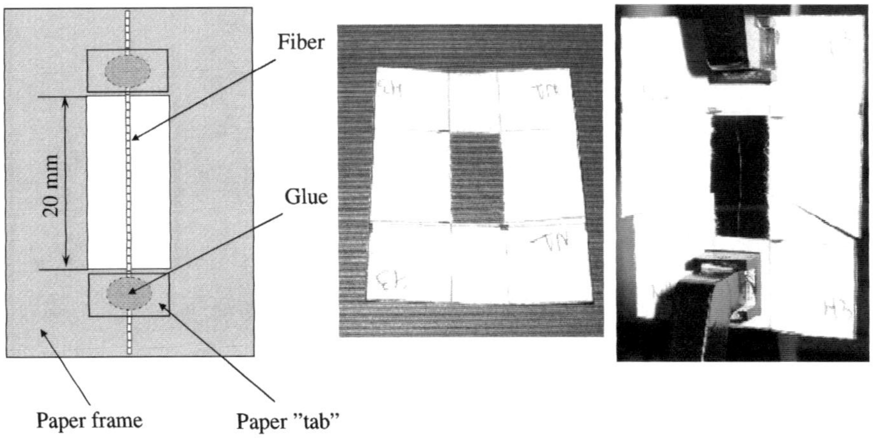

Figure 2. Schematic drawing (left) and photograph (middle) of the single fiber specimen. Photograph of the fiber mounted in the pneumatic grips (paper frame is cut).

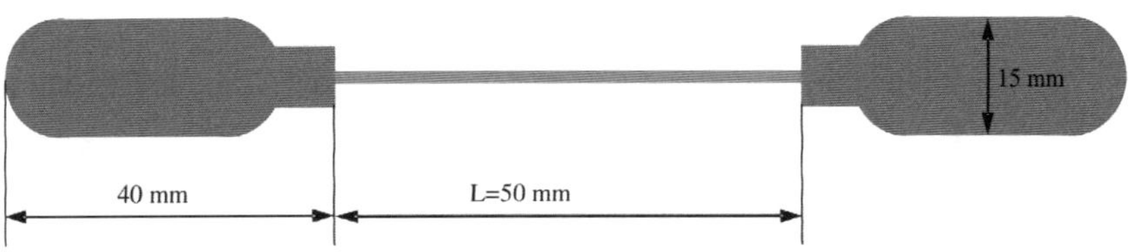

Figure 3. Schematic drawing of the bundle specimen.

2.3.3. Cordenka bundle
Tensile tests of Cordenka bundles were performed on Instron 4411 equipped with 500N load cell and mechanical grips. A schematic representation of the bundle specimen is shown in Figure 3: every bundle is fitted with wooden end tabs by gluing bundle ends in between flat pieces of wood (Araldite 2011 two-component epoxy resin was used as glue).

The average diameter of single Cordenka fiber is d_f=12.5 μm (this value is used to calculate stress). The number of single fibers in the yarn is 1350. The diameter and filament count were determined experimentally and verified from a product data sheet supplied by manufacturer.

2.4. Creep tests

2.4.1. Composite and neat resin
Tensile creep tests were performed using a creep rig with dead weights. In Figure 4 the picture on the top represents the time dependence of the applied stress: creep load is applied in steps and in every step k stress is applied instantly at t_k and kept constant for time interval t_{k+1}-t_k. The picture on the bottom (Figure 4) shows creep strain during loading interval and also strain recovery after load removal. The duration of strain recovery interval was at least 5 times longer than the loading interval. Creep strains (measured using extensometer) were recorded during the loading and also during the following interval of strain recovery after load removal.

The irreversible strains at the end of the recovery period (if present) were analyzed as viscoplastic. Creep tests used for determination of time and load dependence of viscoplastic strains were carried out in multiple steps over shorter time intervals of 10, 20, 30 and 60 min (for composite Cordenka/EpoBioX steps were 3, 7, 15, 30, 60 min).

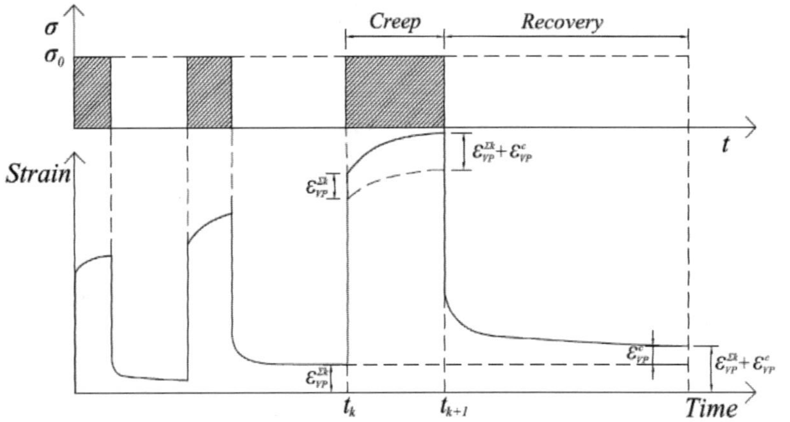

Figure 4. Schematic drawing showing development of strain during creep-recovery test

2.4.2. Cordenka single fiber
Single fiber specimens for creep tests were prepared in the same way as for simple tensile tests, but in this case specimens with a gauge length of 50 mm were prepared. Creep tests were performed on Instron 3366 machine equipped with 5 N load cell and pneumatic grips. Creep tests were performed in one loading step with duration of 2h.

2.5. Accumulation of microdamage and resulting stiffness degradation
After material has been subjected to high stress levels the elastic properties of the material may be degraded. The elastic modulus dependence on the previously applied maximum strain/stress level was measured to evaluate the significance of microdamage developing at high stresses. The strain increment after each step was 0.3%.

3. Results and discussions

It is important to be able to predict long term behaviour of materials under investigation. One of the most general approaches is thermodynamically consistent theory for nonlinear viscoelastic and nonlinear viscoplastic materials developed by Schapery [9,10]. The constitutive equation was further modified to account for microdamage in [11]. The final form of the material model (for one-dimensional case) is

$$\varepsilon = d(\sigma_{\max})\left[\varepsilon_0 + g_1 \int_0^t \Delta S(\psi - \psi') \frac{d(g_2\sigma)}{d\tau} d\tau + \varepsilon_{pl}(t,\sigma)\right] \tag{1}$$

This model has been successfully applied to randomly oriented bio-based composites [12, 13] and long glass fiber composites [14]. This model can be used to predict behaviour of composite as well as constituents (fiber and/or resin). However, this requires a great number of input parameters and therefore a substantial number of experiments must be performed. Considering that a huge variety of different compositions of constituents is possible, a more efficient way to predict the behaviour of composite would be based on constitutive modelling. This means that once constituents (fibers and resins) are fully characterized, the long term properties for resulting composites can be analytically predicted based on this information.

3.1. Mechanical behaviour of regenerated cellulose fibers:

3.1.1. Single fiber tensile tests

Stress-strain curves for single fiber tensile tests are presented in Figure 5 (left). For clarity, only the most representative curves are presented. Average mechanical properties (over all tests) are presented in Table 1. Elastic modulus was measured in strain range of 0.3-0.7%. Presented stress-strain curves and obtained properties show no difference in behaviour of fibers taken out of bundles with or without twist. The stress-strain curve is highly non-linear and further analysis of the shape of the curve shows quasi-linear regions. First linear region is until stress level of 250 MPa, whereas second linear region starts from approximately 350 MPa and continues until failure of the specimen.

Table 1. Average mechanical properties for Cordenka fiber (single fiber tensile test).

Fiber type	Elastic modulus, GPa	Maximum Strain, %	Maximum Strength, MPa
Twisted	21.1±3.2	14.0±3.9	796.8±110.5
Without twist	22.1±2.9	13.5±4.7	769.7±66.5

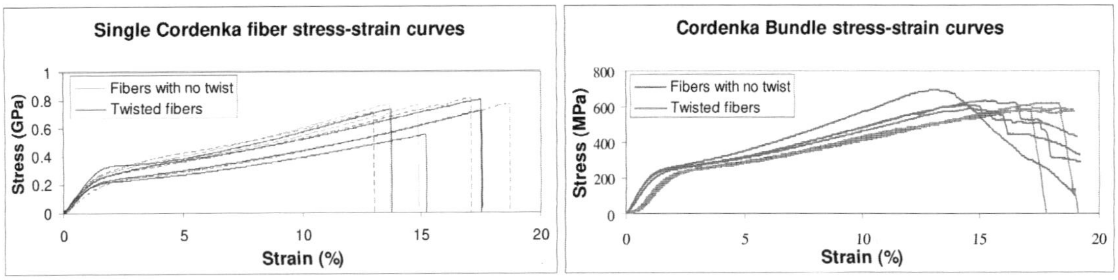

Figure 5. Stress-strain curves (tensile test of single fiber) for regenerated cellulose fibers separated from bundles with and without twist (left) and for regenerated cellulose fiber bundles with and without twist (right).

3.1.2. Fiber bundle tensile tests

Stress-strain curves for Cordenka bundles with and without twist are presented in Figure 5 (right). Average mechanical properties are presented in Table 2. Both types of bundles exhibit similar trends, but stress-strain curves for twisted bundles are slightly shifted to the right on the strain axis. Similarly to single fibers, it is possible to distinguish two linear regions and one transformation region. Stress strain curves for bundles have similar behaviour to single fibers. This is somewhat different from the behaviour of other man-made fibers (e.g. glass and carbon) where the load-displacement curve has a bell-like shape and it can be used to identify parameters of Weibull strength distribution. In the case of synthetic fibers, filaments in the bundle fail progressively and therefore failure of the bundle is not sudden as for Cordenka bundles. This is especially evident for the twisted bundles and less pronounced for the untwisted ones, where the tail of the stress-strain curve looks more similar to the behaviour of synthetic fibers. It can be therefore stated that strength of Cordenka fibers can not be described by the Weibull strength distribution, because these fibers are not brittle material for which this distribution was introduced. Moreover, in the case of twisted bundles fiber are very tightly squeezed together and friction between fibers is too high.

Fibers in twisted bundles are not fully aligned with the direction of loading; therefore they behave similarly to laminates with of-axis fiber orientation which results in lower stiffness in comparison with untwisted bundles. This is evident from the initial part of the stress-strain curve (see Fig. 5). However, at the later stage of tensile test bundles can rotate and fibers start to align with loading direction and thus stress-strain curves for twisted and untwisted bundles are fairly parallel, as seen in Fig. 5.

Table 2. Average mechanical properties for fiber bundles

Bundle type	Elastic modulus, GPa	Strain at maximum stress, %	Maximum Strength, MPa
Twisted	15.68±0.93	18.61±1.94	602.8±17.8
Without twist	23.71±2.51	14.45±0.82	633.2±45.1

The use of fiber bundles with no twist in composites is more preferable than fiber bundles with twist. When fiber bundles with no twist are used, fibers are better dispersed in the matrix. Twisted bundles in composite maintain bundle structure, wheres in composites with bundles with no twist it is not possible to see bundle structure (Figure 6). Fibers in twisted bundle composites have some misaligned angles, thus stresses are not transferred through fiber as efficiently as it would be with completely straight fibers.

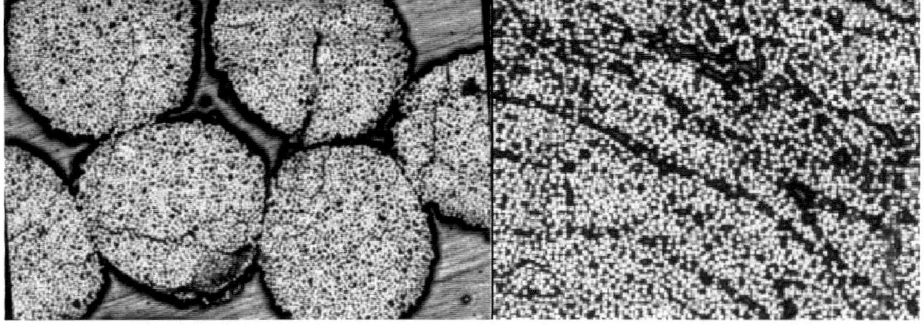

Figure 6. Cross section of Tribest composite with twisted bundles (left) and bundles without twist (right)

3.1.3. Creep tests

Creep tests for single fibers were performed at six different stress levels – 100, 150, 250, 300, 450 and 500 MPa. Creep curves are presented in Figure 7 (left). It is possible to see that regenerated cellulose fibers have significant creep strain even at relatively low stress levels. At 500 MPa, strain reaches approximately 7% and viscoplastic strains 3.5%. These results in combination with results of tensile tests demonstrate that regenerated cellulose fibers exhibit highly visco-elastic and visco-plastic behaviour. Correlation between stress levels in creep tests and two "linearity" regions on the tensile stress-strain curves of fibers and bundles will be addressed in future work.

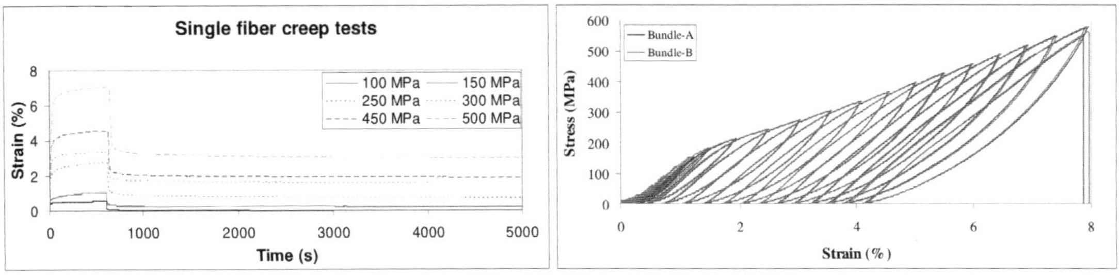

Figure 7. Creep curves for single regenerated cellulose fibers (left) and Stress-strain curves from loading-unloading tests for twisted regenerated cellulose fiber bundles (right)

3.1.4. Loading-unloading tests for regenerated cellulose bundles

Stress-strain curves from loading-unloading tests of regenerated cellulose bundles are presented in Figure 7 (right). Well defined hysteresis loops are clearly visible which is typical phenomena for visco-elastic materials. The overall behaviour of specimens is similar to that in simple tensile tests, but bundles fail at significantly lower strains of 8%, where in simple tensile tests bundles failed around 15%. Stress-strain curves for two tested bundles are almost identical which indicates very good repeatability of experimental results.

3.2. Mechanical behaviour of composites and neat resin

3.2.1. Tensile tests

Stress-strain curves for neat Tribest resin are presented in Figure 8. All experimental curves almost coincide with each other showing excellent repeatability. At the beginning stress-strain curves show linear behaviour, but after approximately stress level of 3 MPa curves have non-linear trend.

Stress strain curves for UD composites are presented in Figure 9 (left) (For more convenient comparison only part of the curves are presented). It is obvious that glass fiber (GF) composites have higher stiffness than regenerated cellulose fiber composites. GF composite curves are linear, whereas regenerated cellulose fiber composite exhibit non-linear behaviour. Comparison of stress-strain curves for GF and Cordenka fiber composites with the same matrix (Tribest) leads to conclusion that non-linear behaviour of composites is mostly defined by fibers. Stress-strain curves for composites with regenerated cellulose fibers but different matrices have similar shape. Composite with EpoBioX has slightly higher modulus. Estimated elastic modulus of neat resin and composites are presented in Table 3.

Figure 9 (right) shows stress-strain curves for Flax and Cordenka fibers UD composites with EpoBioX resin. Flax composites have higher stiffness, but strength for both composites is the same. Both composites exhibit non-linear behaviour, but it is obvious that non-linearity in case of Cordenka composites is more pronounced than for flax composite.

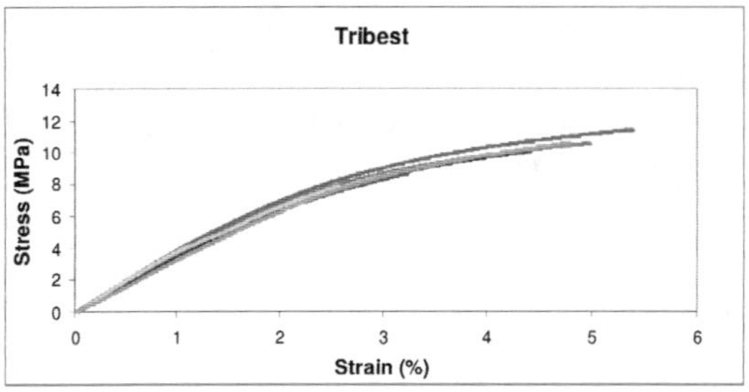

Figure 8. Stress-strain curves for neat tribest resin

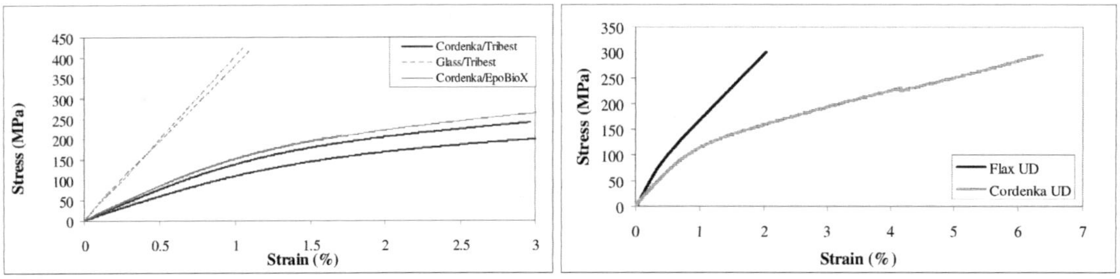

Figure 9. Stress-strain curves for GF and cellulose composites with Tribest and EpoBioX resins (left) and Stress-strain curves for Flax and Cordenka/EpoBioX UD composites (right)

Table 3. Estimated elastic modulus for nest Tribest resin and composites

Material	Elastic modulus, GPa
Neat Tribest resin	0.53
Glass/Tribest UD	43.0
Cordenka/Tribest UD	15.0
Cordenka/EpoBioX UD	17.4
Flax/EpoBioX UD	22.0

3.2.2. Tensile creep tests

Several step creep tests for neat tribest resin were performed at three stress levels 4, 5 and 6 MPa (Figure 10 (left)). Visco-plastic strains for resin at stress levels 4 and 5 MPa are similar. Increase of applied stress by 1MPa (from 5 MPa to 6 MPa) doubles level of visco-plastic strain.

Several step creep test for Cordenka/EpoBioX composite are presented in Figure 10 (right). The composite failed at 130 MPa in creep. While in tensile test, composite failed approximately at 470 MPa. This means that composite fails in creep at 3.6 times lower stress than in tensile test, which certainly depends on the strain rate in the tensile test and duration of the creep test. It also indicates that material exhibits highly visco-elastic and visco-plastic behaviour.

Several step creep test for GF/Tribest composite is presented in Figure 11 (left). As expected, GF composite exhibits linear elastic behaviour also in creep. There are no significant visco-elastic strains. At higher stress levels there are very small visco-plastic strains.

Figure 11 (right) shows several step creep test for neat Tribest resin at 6 MPa and composites with GF and regenerated cellulose fibers at 90 MPa. The strain for composites with GF is less than half of that for composites with Cordenka. However, it should be noted that even though both, fiber and matrix, in Cordenka fiber composites is susceptible to significant creep when tested separately, composite does not exhibit extreme creep when tested at low stresses (when compared to the failure stress).

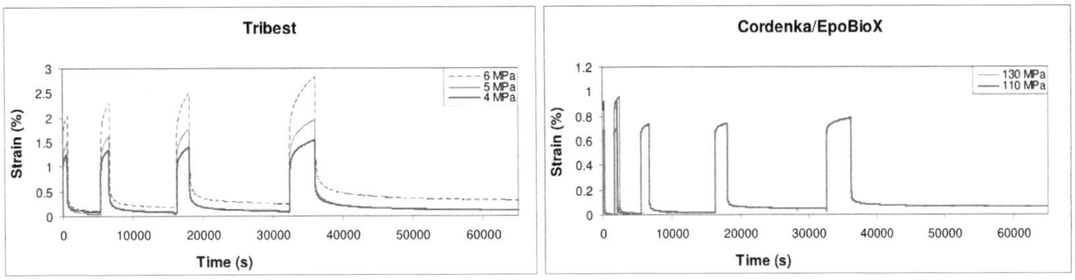

Figure 10. Several step creep test results for neat tribest resin (left) and Cordenk/EpoBioX (right)

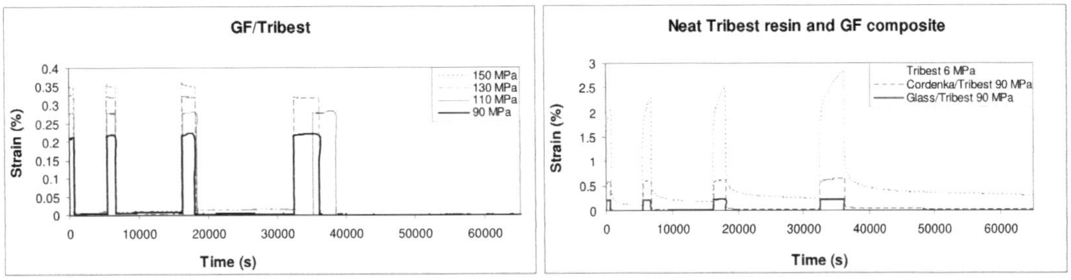

Figure 11. Several step creep tests results for GF/Tribest composite (left) and neat Tribest resin at stress level 6 MPa and Cordenka/Tribest and GF/Tribest composites at 90 MPa

4. Summary

The highly non-linear behaviour of Cordenka fibers was demonstrated via tensile and creep tests of bundles and monofilaments. At high stresses (>300 MPa) Cordenka fibers exhibit very significant creep, whereas at low applied stress (<100 MPa) evolution of strain during the creep test is not excessive and almost no residual strain develops.

Two types of bundles, with and without twist, were tested. The twisting of fiber bundles does not seem to influence properties of elementary fibers. The fibers extracted from bundles with or without twist shows similar properties suggesting that in the twisting process fibers are not considerably damaged. However, bundles without twist perform slightly better than those with twist. Moreover, use of untwisted bundles is mush more beneficial for manufacturing of composites: fibers are dispersed more homogeneously throughout the material. In tension the bundles of Cordenka behave qualitatively likewise to the single fibers, shape of stress-strain curves is similar, this is very different from the behaviour of other man made fibers (e.g. glass, carbon fibers). But quantitatively there are some differences in mechanical performance: single fibers show higher strength than bundles (780 MPa vs 620 MPa), whereas the strain at maximum stress is higher for bundles than for single fibers (14% vs >15%).

Tensile tests of composites show that GF/Tribest behaves linearly with elastic modulus of 43 MPa. It should be noted that this value is rather close to the stiffness of commonly used UD prepreg GF/Epoxy composite with high volume content of fibers (value of 45-47 GPa has been reported). But the composites made out of flax and Cordenka fibers with Tribest resin behave non-linearly and show much lower stiffness. However, stiffness of Cordenka fibers is 3.6 times lower than that of GF but stiffness of Cordenka fibers composites is only 2.9 lower than elastic modulus of GF composites. This means (assuming similar volume fractions of fibers in GF and Cordenka laminates) that properties of Cordenka are well utilized due to homogeneous fiber distribution and well controlled fiber orientation in the material. There is small difference between stiffness of Tribest and EpoBioX resin based composites (with Cordenka fibers): EpoBioX composite has 15% higher modulus than composite with Tribest. This of course can be explained by difference in properties of matrix (stiffness of Tribest is <1GPa, whereas stiffness of EpoBioX is 2.5GPa) and possible small deviations in fiber content.

Neat Tribest resin showed high visco-elastic and visco-plastic response in creep tests, whereas GF/Tribest composites did not creep at all. The strain developed in GF composite during the creep were 2.5 times smaller than for Cordenka composites but it can be explained by difference in stiffness. In general, it can be stated that at small applied stresses (<100MPa) Cordenka composite does not exhibit extreme creep.

The results accumulated in this investigation facilitate understanding of behaviour of composites with highly non-linear fibers. It will promote development of these materials as well as help in validation of models for prediction of performance of such composites. This knowledge will be further used to design materials with new compositions (and properties) tailored according to the particular applications.

Acknowledgment
Part of this study was financially supported by Interreg IVA Nord project ANACOMPO (funding by EC and regional government of Norrbotten). Manufacturing of composites was financially supported by VINNOVA/project Nanofibre.

Authors would like to acknowledge Fanny ROOS (EEIGM) for performing part of experiments and Runar Långström (Swerea SICOMP) for composite manufacturing.

References
[1] Oksman K 2000 Liquid Composite Moulding of Natural fibres *Natural Agrofibres based composites* (Sao Carlos: Embrapa Instrumentacao Agropecuaria) p 203-227
[2] Goutianos S, Peijs T, Skrifvars M and Eng T 2003 Textile reinforcements based on flax fibres for structural composites. Part 1 – The optimisation of flax fibre yarns for high-performance natural fibre composites. *Poster presentation at the 2nd Int. Conf. on Eco-Composites* (London) p 3.04
[3] Gea S, Goutianos S, Peijs T, Skrifvars M and Eng T 2003 Textile reinforcements based on flax fibres for structural composites. Part 1 – Mechanical properties of fabric based laminates. *Poster presentation at the 2nd Int. Conf. on Eco-Composites (London)* p3.04
[4] Flodin P and Zarodecki P 1986 *Composite systems from Natural and Synthetic Polymers* (Amsterdam: Elsevier) p 59
[5] Wool R and Sun X S 2005 *Bio-based polymers and composites* (Burlington: Elsevier)
[6] Mohanty A K, Misra M and Drazal L T 2008 *Natural fibers, biopolymers and biocomposites* (Broken Sound Parkway NW: Taylor and Francis Group LLC)
[7] Pickering KL (2008) *Properties and performance of natural-fibre composites* (Broken Sound Parkway NW: Woodhead Publishing)
[8] "Standard Test Method for Tensile Strength and Young's Modulus for High-Modulus Single Filament Materiala", ASTM D3379-75 (Reapproved 1989)
[9] Lou Y C and Schapery R A 1971 Viscoelastic characterization of a nonlinear fiber-reinforced plastic *J. Compos. Mater.* **5** 208-34
[10] Schapery R A 1997 Nonlinear viscoelastic and viscoplastic vonstitutive equations based on thermodynamics *Mech. Of time dependent mater.* **1** 209-40
[11] Marklund E, Eitzenberger J and Varna J 2008 Nonlinear viscoelastic viscoplastic material model including stiffness degradation for hemp/lignin composites *Compos. Sci. Tech.* **68** 2156-62
[12] Varna J, Rozite L, Joffe R and Pupurs A 2011 Nonlinear behavior of PLA based flax composites *Plast. Rubber Compos.* accepted
[13] Sparnins E, Varna J, Joffe R, Nättinen K and Lampinen J 2011 Time dependent behavior of flax/starch composites *Mech. of Time-Dependent Mater.* Accepted
[14] Giannadakis K, Mannberg P, Joffe R and Varna J 2011 The sources of inelastic behaviour of glass fiber/vinylester non-crimp fabric $[\pm 45]_s$ laminates *J. Reinforc. Plast Compos.* Available on-line prior to publication

IOP Conf. Series: Materials Science and Engineering **31** (2012) 012006 doi:10.1088/1757-899X/31/1/012006

Characterization of cellulose nanowhiskers: A comparison of two industrial bio-residues

M A. Herrera, A P. Mathew and K Oksman
Department of Engineering Sciences and Mathematics
Luleå University of Technology, SE-97187 Luleå, Sweden

E-mail: martha.herrera@ltu.se

Abstract. Cellulose nanowhiskers separated from two different industrial residues, sludge from cellulose production (CNW_{SL}) and lignin residue from ethanol production (CNW_{ER}), were compared in order to evaluate their characteristics and their potential as a source for the production of cellulose nanowhiskers (CNWs). It was found that CNW_{SL} and CNW_{ER} suspensions exhibited flow birefringence when they were studied through cross-polarized filters. Transmission electron microscopy (TEM) study showed that the CNW_{SL} were longer (377 nm) than CNW_{ER} (301 nm). It was also demonstrated that most CNW_{SL} had nanowhiskers between 375-449 nm and CNW_{ER} between 300-374 nm. The UV/Vis spectroscopy showed a stronger interference in the UV and visible region for the CNW_{SL} films. The crystallinity, obtained by X-ray analysis, was higher for CNW_{SL} (86%) than for CNW_{ER} (78%). Finally, the thermal stability appeared to be slightly higher for the CNW_{ER} than for CNW_{SL}. Both studied residues seem to be suitable sources for large-scale production of CNWs.

1. Introduction

Cellulose is the structural component of the cell wall of green plants and some algae. This makes it one of the most abundant and naturally occurring polymers on earth. Due to its availability, along with its sustainability and abundance, cellulose has been widely used in the manufacture of paper, fibers and films [1,2]. A cellulose fiber is a bundle composed of several cellulose nanofibers. These nanofibers have an amorphous part and a crystalline part known as cellulose nanowhiskers (CNWs). Even though cellulose is abundant, CNWs are not yet commercially available. CNWs are rod-shaped nano-sized crystals which can be obtained by acid hydrolysis from different natural sources. These nanocrystals have a high elastic modulus (120-140 GPa) [3], very close to the theoretical value (167.5 GPa) [3], and therefore have been extensively used as a reinforcement material in biopolymers [4]. Depending on the source, the sizes of the CNWs can vary but, in general, the data in Table 1 can be used as a reference [3, 4, 5]

Table 1: CNWs general properties

Property	Cellulose nanowhiskers
Length (nm) [4]	100-1000
Diameter (nm) [4, 5]	4-25
Aspect Ratio (L/d) [4]	10-100
Elastic Modulus (GPa) [3]	120-140

Our group has focused during recent years on studying residue from forest-based industries as a source of material for the production of cellulose nanowhiskers. In 2011, a residual material from wood ethanol production was used for the separation of the cellulose nanocrystals [6]. Oksman et al. demonstrated that this residue, usually called lignin residue, consists of almost 50% cellulose and this cellulose was shown to be highly crystalline, because the material has been subjected to strong acid hydrolysis steps during the ethanol production process [6]. The CNWs from this source were obtained with only purification and homogenization steps, obtaining a yield of 43% [6]. The study showed that lignin residue is a viable source for the production of cellulose nanowhiskers.
The aim of this work was to examine the characteristics of CNWs obtained from two different industrial residues. For this purpose, flow birefringence, morphology, size, transparency, comparative crystallinity and thermal stability of nanowhiskers extracted by sulfuric acid hydrolysis from sludge (CNW_{SL}) and homogenized whiskers from bioethanol residue (CNW_{ER}) were compared.

2. Experimental Methods

2.1 Materials

Cellulose nanowhiskers suspensions were extracted from two different raw material sources. The first one was a sludge from cellulose production from Domsjö Fabrikerna AB, Örnsköldsvik, Sweden. The nanowhiskers (CNW_{SL}) were extracted by acid hydrolysis following the procedure reported by Bondeson et al. in 2006 [7] with minor difference. The second source used was a residue of wood ethanol production from SEKAB E-Technology, Örnsköldsvik, Sweden. These nanowhiskers (CNW_{ER}) were extracted using the procedure described by Oksman et al. in 2011 [6]. The schematic representation of the isolation routes adopted for the studied materials is given in figure 1.

Figure 1: Scheme for the CNWs extraction from the sources used.

2.2 CNW films

Thin films of between 20 and 40 nm were prepared from the whiskers suspensions. 40 mL of a 2% suspension was casted and dried under ambient conditions for approximately 12 h. These films were used for further characterizations.

2.3 Flow birefringence

The flow birefringence behavior of the CNWs suspensions was studied using a set consisting of 2 cross-polarized filters and a lamp.

2.4 Nanostructure
An HT 7700 120 kV Compact-Digital biological transmission electron microscope (Hitachi, Tokyo, Japan) was used at a voltage of 100 kV. For this purpose, a drop of a 0.2 wt % CNWs dispersion was allowed to dry on a copper grid covered with porous carbon film. The whiskers size determination was done with the digital slow scan image recording system software SemAfore 4.01 JEOL (Skandinaviska) AB (JEOL, Sweden).
The measurement of the length distribution was done manually form the TEM images. At least 100 length measurements from 4 different images for each sample were performed.

2.5 UV/VIS spectroscopy
Prepared nanowhiskers films were tested in a Perkin Elmer UV/Vis spectrometer Lambda 2S (Perkin Elmer, Überlingen, Germany) to compare their transmittance patterns. The wavelength used was from 700 nm to 300 nm with a scan speed of 240 nm/min and a data collection interval of 0.50 nm. The reported data for each material represent the average of three samples.

2.6 X-ray diffraction
X-ray diffraction was used to compare the crystallinity of the materials. A film of each CNW was analyzed at ambient temperature by step scanning on an X-ray diffractometer Siemens D5000 (Siemens, Berlin, Germany) using a monochromatic CuKα radiation (λ=1.54 Å) in the range of 2θ=5°-38° with a step size of 0.02° and a scanning rate of 0.5°/min. The crystallinity index (C_{Ir}) was measured using the Segal empirical formula [8]:

$$C_{Ir} = [(I_{200} - I_{am}) / I_{200}] * 100\% \quad (2.1)$$

Where I_{200} is the intensity value for the crystalline cellulose, and I_{am} is the intensity value for the amorphous cellulose. The data presented are the average, with standard deviation, of four measurements for each kind of sample.

2.7 Thermal stability
The thermogravimetric analysis (TGA) was carried out for the CNW_{SL}, and CNW_{ER} in a TGA Q500 (TA Instruments, New Jersey, U.S.A.). The temperature range was from 60°C to 510°C in an air atmosphere, with a heating velocity of 5°C/min. The data presented are the average of three measurements for each material.

3. Results and discussions

3.1 Flow birefringence
Figure 2 shows the flow birefringence for the CNW_{SL} and CNW_{ER} suspensions. Both suspensions had the same concentration of whiskers (0.1%), and flow birefringence could be seen. This behavior demonstrates the fact that rod- shaped cellulose was obtained in the extraction process [9].

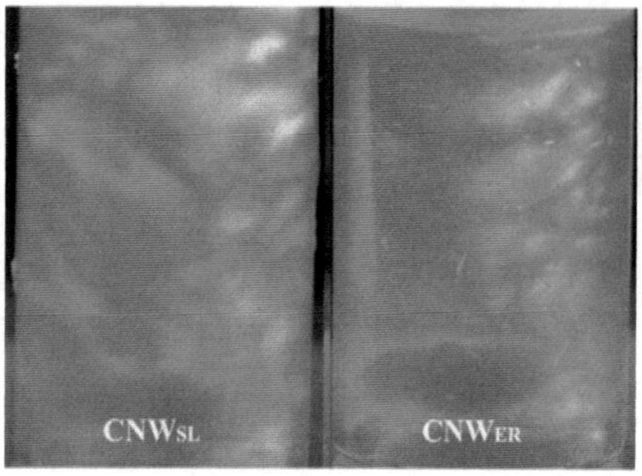

Figure 2: Flow birefringence of CNW_{SL}, and CNW_{ER}.

3.2 Nanostructure

Transmission electron microscopy (TEM) was used to observe the morphology and measure the lengths of the extracted nanowhiskers. The images obtained from the TEM are shown in figure 3, where it can be seen that both whiskers appear to be similar in shape. It was observed that both materials presented agglomeration that followed a parallel arrangement of the whiskers. Similar characteristics were reported earlier by other authors [10, 11].

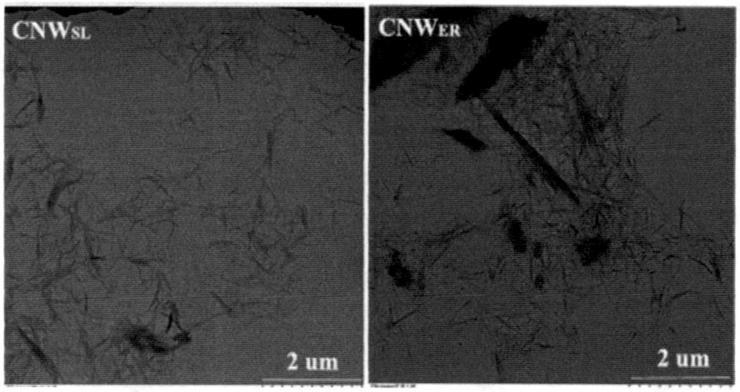

Figure 3: TEM images of CNW_{SL} and CNW_{ER}.

The length values are listed in table 2 and the length distribution is shown in figure 4. The accuracy of length measurement depends on the resolution of the images; therefore, there is a risk that the whiskers ends are not included in the measurement. Also, the agglomeration seen in the figure 3 causes difficulties in the manual measurement. For this reason, the lengths represented in table 2 and figure 4 are given only for comparison purposes.

Table 2: Data obtained from the different tests on the CNW_{SL} and CNW_{ER}.

Source	Length (nm)	Crystallinity (%)	Degradation onset temperature (T_0) (°C)	Peak degradation temperature (T_{max}) (°C)
CNW_{SL}	377 ± 132	85.8 ± 4.6	198.2 ± 20.7	287 ± 5
CNW_{ER}	301 ± 126	77.7 ± 7.0	218.5 ± 6.0	296 ± 21

In table 2 it can be seen that CNW_{SL} are slightly longer than CNW_{ER}. Despite the information obtained for the average length of the whiskers, it was also interesting to measure the length distribution of the whiskers (figure 4). It was observed that the lengths of CNW_{SL} were in the range of 375-449 nm and most of the CNW_{ER} were between 300-374 nm.

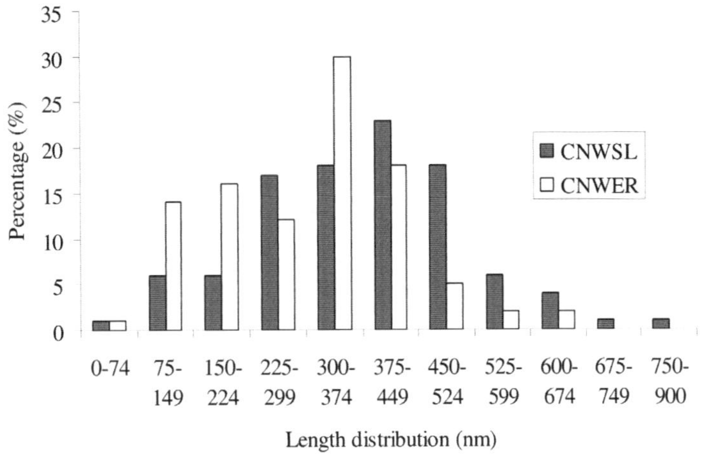

Figure 4: Length distribution for the CNW_{SL} and CNW_{ER}.

The differences in the length distribution can be attributed to the differences in the extraction procedures of the materials. The indication that CNW_{SL} are longer than CNW_{ER} might depend on the extraction procedure used for the CNW_{SL}, which may have been less detrimental to the crystals than that used for CNW_{ER}. An earlier study by Bondeson et al. [7] showed that longer hydrolysis time had a negative effect on the size of the cellulose nanocrystals.

3.3 UV/Vis spectroscopy

UV/Vis spectroscopy was used to measure the differences in light absorption of the films. The wavelength applied was in the range from near ultraviolet (300-400 nm) to visible light (400-700 nm). The results are shown in Figure 5. In the range of visible light, the CNWs films seem to have a stable tendency towards decreased transmittance when the wavelength is decreasing as well. The CNW_{SL} films demonstrated higher interference when compared with CNW_{ER}. In figure 5 it could be seen that none of the films permitted a complete transmission of light through them, the maximum value being 60% transmittance for the CNW_{ER} films..

When the materials are exposed to UV light, an interesting behavior appears in the CNW_{ER} films. These films show a faster decrease in transmittance values when compared to the CNW_{SL} films. This difference in the tendency, in the CNW_{ER} films, might be caused by the differences in the length of the whiskers seen in table 2. Particles with lengths around 300-400 nm may have caused interference in this wavelength range, showing a resultant decrease in transmittance. This could be translated into a stronger interference in the UV range. Although this test has not been used previously for the analysis of films made with pure cellulose, it has been applied to polymer/CNWs composite films. It was demonstrated that the agglomeration of CNWs can cause interference in the lower spectrum, causing a decrease in transmittance of UV light [11, 12]. If the CNW_{ER} films are not completely homogeneous, regions of high cellulose nanowhiskers density can be analogues to the agglomerated regions in the polymer/CNWs composites, producing a similar interference effect.

3.4 X-ray diffraction

X-ray diffraction was used to characterize the crystallinity and the specific polymorphism of the CNWs. Previous studies have demonstrated that celluloses obtained from different sources and different acid hydrolysis (sulfuric and hydrochloric) have similar XRD patterns. It is also shown that, when cellulose I is present, there are three characteristic peaks in the diffractograms. These are two shoulder-like peaks around 14° and 16°, and a narrower intense peak at around 22° [2, 6, 7].

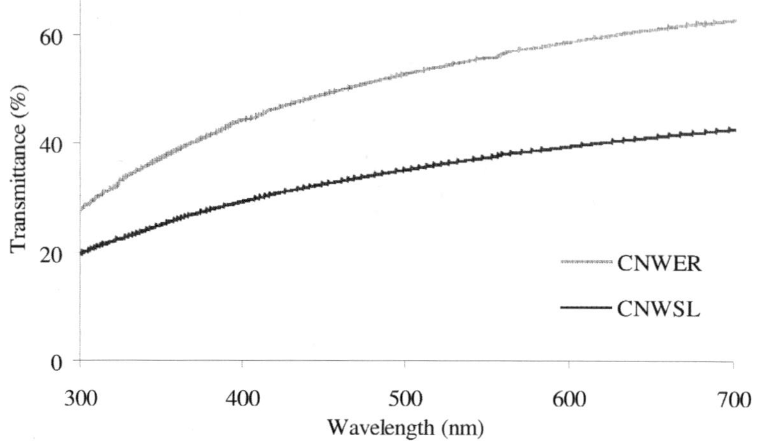

Figure 5: UV/Vis spectroscopy patterns for CNW_{ER} and CNW_{SL}.

The information obtained from the X-ray diffraction of CNWs is given in Figure 6 and table 2. In the diffractograms, in Figure 6, two well-defined parts can be observed, a shoulder around $2\theta=15°$ and a very narrow peak at $2\theta=22.5°$. Those parts correspond to the cellulose I structure [1].The values of the crystallinity for both materials are listed in table 2. The analysis showed that the CNW_{SL} had slightly higher crystallinity, 86%, compared to that of the CNW_{ER}, 78%. Determination of the crystallinity by the Segal method is widely used [6, 14, 15], but it is not an accurate method and, therefore, the percentage of crystallinity in this study should be considered more as comparative rather than absolute values.

3.5 Thermal stability

The thermogravimetric analyses of the whisker films are represented in Figure 7. The onset degradation (obtained by the TGA analysis) and the peak degradation temperatures (obtained by the DTG analysis) are listed also in the table 2. The CNW_{SL} presented a lower degradation onset and peak degradation temperature when compared with the CNW_{ER}. Of the two, CNW_{ER} demonstrated a better thermal stability.

During the acid hydrolysis, with sulfuric acid, an introduction of sulfate groups in the whiskers surface occurs. These groups make the CNWs suspension a stable colloidal system, due to the electrically negative charge on the surface [4, 11]. The sulfate groups also affect the thermal stability of the whiskers, due to a dehydration reaction, making them more susceptible to degradation at elevated temperatures [16, 17]. This explains why CNW_{SL} is less thermally stable and less agglomerated than $CNW_{ER.}$ Even though CNW_{ER} was obtained only by mechanical procedures (homogenization); the residue material passed through a two-step acid hydrolysis when the ethanol extraction process was done. However, after that, the ER was subjected to a dewaxing, bleaching, cleaning and homogenization processes [6] in order to extract the cellulose nanowhiskers from it. These steps seem to have cleaned the whiskers surface from these sulfate groups, restoring its thermal stability. Another proof of this loss of sulfate groups on the CNW_{ER} surface is the fact that the agglomeration for these whiskers was higher than for CNW_{SL}, as seen in figure 3.

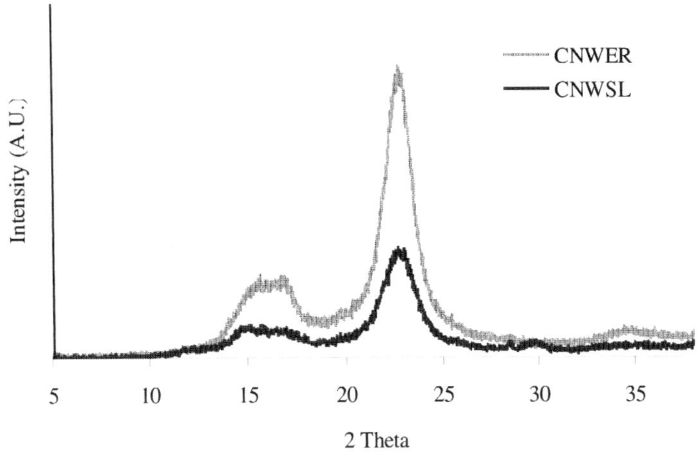

Figure 6: X-ray diffraction patterns for the cellulose whiskers extracted from sludge and ethanol residue.

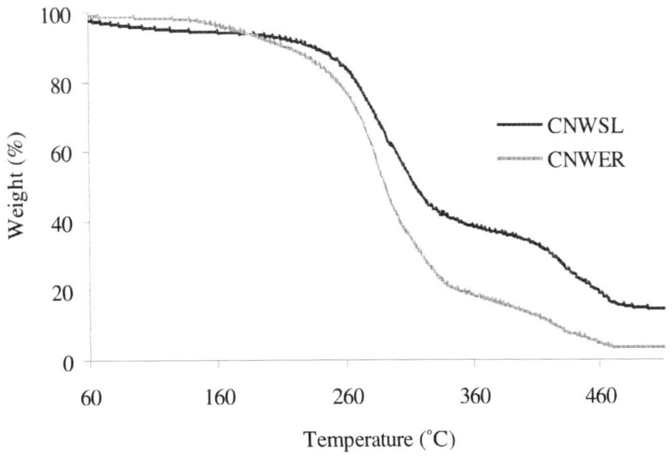

Figure 7: Thermal stability of the cellulose nanowhiskers from the sludge (CNW_{SL}) and ethanol residue (CNW_{ER}).

4. Conclusions

The properties of cellulose nanowhiskers separated from two different industrial residues, sludge from cellulose production and lignin residue from wood bioethanol production, were compared. The procedure for separation of nanowhiskers from these sources also differed. Sulphuric acid hydrolysis was used for the sludge and bleaching and high-pressure homogenization was used for the lignin residue.

Flow birefringence was observed in both samples, indicating that the whiskers are well individualized. The transmission electron microscopy study showed that the length distribution for CNW_{SL} was in the range of 375-449 nm, and that the CNW_{ER} were slightly shorter, having a length distribution between 300-374 nm.

The UV/Vis spectroscopy revealed that the films were not transparent in the UV and visual spectrum, and that the CNW_{SL} displayed a higher interference in all the ranges studied, confirming the presence of longer crystals observed in the TEM study.
The results from the X-ray diffraction analysis showed that both materials displayed cellulose I structure. The crystallinity of whiskers extracted from sludge was approx. 86% and for the CNWs from ethanol residue was somewhat lower, 78%.
The TGA study demonstrated that the whiskers extracted from ethanol residue were more thermally stable than those extracted from sludge. The reason might be that several washing steps were used in the extraction of the CNW_{ER}.
This work demonstrates that sludge from cellulose production and lignin residue from wood bioethanol production can potentially be used as raw materials to produce value-added products from bio-residues, thereby increasing the value of the forest resources.

5. References

[1] Krässing H A, 1996 Cellulose. Structure, accessibility and reactivity. Volume11. (Netherlands, Gordon and Breanch Science Publisher). Chapter 1
[2] Taniguchi T 1998. Polym. Inter. 47 291-294.
[3] Tashiro K 1991 Polym. 32 1516-1526.
[4] Favier V 1995 Macromolecules. 28 6365-6367.
[5] de Rodriguez N, Thielemans W, Dufresne A, de Rodriguez G 2006 Cellulose. 13 261-270.
[6] Oksman K, Etang J, Mathew A, Jonoobi M 2011 Biomass Bioenerg. 35 146-152.
[7] Bondeson D, Mathew A, Oksman K 2006 Cellulose. 13 171-180.
[8] Segal L, Creely L, Martin A E, Conrad C M 1959 Text Res. J. 29 786-794.
[9] Oksman K, Mathew A, Sain M. 2009 Plastic, Rubber and Composites. 38 396-405.
[10] Kvien I, Tanem B, Oksman K. 2005 Biomacromolecules. 6 3160-3165.
[11] Marchessault R H 1959 Nature. 184 632-633.
[12] Petersson L, Oksman K. 2006 ACS symposium series 938 132-150.
[13] Petersson L, Mathew A, Oksman K. 2009 J. Appl. Polym. Sci. 112 2001-2009.
[14] Filson P, Dawson-Andoh B. 2009 Bioresour. Technol. 100 2259-2264.
[15] Siqueira G, Bras J, Dufresne A 2010 Langmuir. 26 402-411.
[16] Roman M, Winter W 2004 Biomacromolecules. 5 1671-1677.
[17] Rosa M, Medeiros E, Malmonge J, Gregorski K, Wood D, Mattoso L 2010 Carbohydr. Polym. 81 83-92.
[18] Dong X, Revol J, Gray D. 1998 Cellulose. 5 19-32.

Acknowledgements
The authors would like to thank Bio4Energy for financial support, and SEKAB E-Technology in Örnsköldsvik, Sweden and Domsjö Fabrikerna AB, Örnsköldsvik, Sweden for supplying the bio-residues. The authors also wish to thank Dr. M. Jonoobi for his help in the realization of the TEM images presented in this paper.

IOP Conf. Series: Materials Science and Engineering **31** (2012) 012007 doi:10.1088/1757-899X/31/1/012007

Degradation of chitosan-based materials after different sterilization treatments

A San Juan[1], A Montembault[1*], D Gillet[2], J P Say[3], S Rouif[4], T Bouet[5], I Royaud[1,6] and L David[1]

[1]Université de Lyon, F-69361, Lyon, France; CNRS, UMR 5223, Ingénierie des Matériaux Polymères, F-69622, Villeurbanne, France; Université Claude Bernard Lyon 1, F-69622, Villeurbanne, France
[2]Gillet Chitosan, 17 Rue Laurent Bonnevay, F-54100 Nancy, France
[3]France Chitine, Chemin de Porte Claire, F-84100 Orange, France
[4]Ionisos, Parc Dombes Cotière Activités, F-01120 Dagneux, France
[5]Laboratoire Tetra Medical, PAE de Marenton, F-07104 Annonay, France
[6]Institut Jean Lamour, UMR CNRS 7198, Nancy Université, Département SI2M Ecole des Mines- Parc de Saurupt- CS 14234 ; 54 042 Nancy Cx, France

*Corresponding author : A. Montembault, Tel : + 33 472431604; Fax: + 33 478892583
e-mail: alexandra.montembault@univ-lyon1.fr

Abstract. Biopolymers have received in recent years an increasing interest for their potential applications in the field of biomedical engineering. Among the natural polymers that have been experimented, chitosan is probably the most promising in view of its exceptional biological properties. Several techniques may be employed to sterilize chitosan-based materials. The aim of our study was to compare the effect of common sterilization treatments on the degradation of chitosan-based materials in various physical states: solutions, hydrogels and solid flakes. Four sterilization methods were compared: gamma irradiation, beta irradiation, exposure to ethylene oxide and saturated water steam sterilization (autoclaving). Exposure to gamma or beta irradiation was shown to induce an important degradation of chitosan, regardless of its physical state. The chemical structure of chitosan flakes was preserved after ethylene oxide sterilization, but this technique has a limited use for materials in the dry state. Saturated water steam sterilization of chitosan solutions led to an important depolymerization. Nevertheless, steam sterilization of chitosan flakes bagged or dispersed in water was found to preserve better the molecular weight of the polymer. Hence, the sterilization of chitosan flakes dispersed in water would represent an alternative step for the preparation of sterilized chitosan solutions. Alternatively, autoclaving chitosan physical hydrogels did not significantly modify the macromolecular structure of the polymer. Thus, this method is one of the most convenient procedures for the sterilization of physical chitosan hydrogels after their preparation.

1. Introduction

Among the natural polymers that have been experimented for potential applications in the field of biomedical engineering, such as drug delivery, wound healing, and tissue engineering, chitosan has been one of the most extensively studied. This linear polysaccharide of D-glucosamine and N-acetyl-D-glucosamine units linked by 1-4 glycosidic bonds is the second most abundant polysaccharide in nature. Chitosan is well known for its biocompatibility [1], hemostatic, bacteriostatic and fungistatic properties, bioresorbability as well as its bioactivity [2]. For several years, the authors have been developing chitosan-based materials for various biomedical applications [3,4,5]. As any medical device implanted in human body, these materials must be sterilized prior to use.

Common methods for the sterilization of medical devices include treatments using dry heat, saturated water steam, ethylene oxide (EtO) and ionizing gamma or beta radiations. The action of these sterilization techniques is based on physical or chemical disabling or destruction of microorganisms. Depending on their nature, sterilization processes can also change the macromolecular structure of polymers by hydrolysis, oxidation, chain scission, and depolymerization [6].

The aim of the present study was to compare ionizing radiations, EtO and saturated water steam sterilization treatments for the preparation of sterilized pure chitosan-based materials in different physical forms, namely: dry materials (flakes, films, fibers), solutions and hydrogel-type materials (monolithic gels, tubes). Thus, in this work, the effects of these sterilization treatments on the macromolecular structure of chitosan were compared for chitosan flakes, solutions and hydrogels.

2. Experimental details

2.1. Materials

Samples of chitosan flakes, derived from squid pens, were supplied by France Chitine and used as received. The degree of acetylation (DA) of chitosan corresponds to the molar fraction of acetyl units within the polymer chains. A batch of chitosan was reacetylated [7] to obtain a sample of higher degree of acetylation (35%) with a statistical distribution of the residues. The DA of the tested chitosan samples ranged between 3% and 35%, the weight-average molecular weights obtained were close to 500 000 g/mol, with an average polydispersity index of 1.7.

2.2. Preparation of chitosan solutions

Chitosan was dispersed in de-ionized water. Acetic acid was added to achieve the stoechiometric protonation of the -NH_2 sites. Then, the system was kept under magnetic stirring at room temperature until complete dissolution of chitosan (minimum 12 h of stirring). The polymer concentration of the studied solutions was 1.5% (w/w). The pH of solutions was found around 4.5.

2.3. Preparation of chitosan hydrogels

Two types of physical chitosan hydrogels were prepared from chitosan solutions as described in the previous paragraph but with an initial polymer concentration of 3% (w/w). Hydrogels were obtained either from an aqueous chitosan solution which was put in contact with gaseous ammonia to induce gelation, or from a hydro-alcoholic solution, as previously described by Montembault *et al.* [7, 8].

2.4. Gamma irradiation

Gamma irradiation was performed by IONISOS (01120, Dagneux, France). Samples were gamma-irradiated using a ^{60}Co source at a dose of 15 or 40 kGy with an average dose rate of 2-3 kGy/h (the standard sterilizing dose being 25 kGy).

2.5. Beta irradiation

Beta irradiation was performed by IONISOS (10500, Chaumesnil, France). Samples were beta-irradiated at a dose of 15, 25 or 40 kGy with a dose rate of about 300 kGy/min (beam of 10 MeV-energy and of 20 kW-power). The standard sterilizing dose is also 25 kGy.

2.6. Ethylene oxide

EtO sterilization of chitosan flakes was performed by Tetra Medical Laboratory (07104, Annonay, France). The different stages of a sterilization cycle can be described as follows: pre-conditioning at 50°C with 75% humidity during 4 h, storage between the pre-conditioning room and the autoclave, autoclaving at 50°C with 75% humidity in a gaseous mixture of EtO and nitrogen during 4 h, with a pressure decrease of 900 mbars followed by at least 3 stages of desorption (vacuum/atmosphere).

2.7. Characterization of chitosan

The degree of acetylation (DA) of chitosan was determined by proton NMR. ^{1}H-NMR spectra were obtained with a Bruker 250 (250 MHz) at 25°C. 10 mg of chitosan were solubilized in 1g of D_2O containing 0.21% by weight of HCl. The DA was obtained according to the method described by Hirai *et al* [9].

The weight-average molecular weight (Mw) and polydispersity index (Ip) of chitosan were determined by multi-angle laser light scattering combined with size exclusion chromatography on polymeric solutions at a concentration of 1 mg/mL in ammonium acetate 0.15 M / acetic acid 0.2 M (pH 4.5), according to a protocol already described [7].

3. Results and Discussion

The effects of gamma irradiation, beta irradiation, EtO exposure, and saturated water steam sterilizations on the macromolecular structure of chitosan were studied in the case of different types of materials: chitosan flakes, chitosan in solution in diluted acid medium or in a physically cross-linked hydrogel.

Ionizing irradiations are often employed for the sterilization of heat sensitive pharmaceuticals and devices for medical applications, because they allow easier packaging and relatively low processing temperatures. In the present study, gamma irradiation was shown to drastically affect the structure of chitosan-based materials, as revealed by the important decrease in molecular weight observed for all types of exposed samples as soon as the 15 kGy dose was received (see table 1). For chitosan flakes and hydrogels, the extent of chitosan degradation (evaluated as a decrease of Mw) increased with increasing the irradiation dose. In the case of chitosan solutions, a drastic degradation was already observed from 15 kGy but it did not increase significantly for higher irradiation doses. Beta irradiation of chitosan-based materials also resulted in a strong depolymerization of chitosan in all physical states (data not shown), in the same range that was observed for gamma irradiation. Chitosan hydrogels which had undergone such ionizing sterilization processes were found broken into several parts.

Table 1. Decrease of Mw (in %) after gamma irradiation sterilization of different chitosan based materials.

Gamma irradiation dose	15 kGy	40 kGy
Flakes	70 %	85 %
Solutions	98 %	98 %
Hydrogels	80 %	90 %

NMR analysis of chitosan flakes after gamma or beta irradiation did not show any significant chemical change. No significant modification of the DA was observed. However as gamma and beta irradiation strongly decreased the chitosan molecular weight, regardless of the physical form of the polymer, these sterilisation treatments should not be recommended for the sterilization of chitosan-based materials, unless the purpose is also to reduce the molecular weight of chitosan in order to accelerate the bioresorption or enrich the material with low molecular weight chitosan in order to favor tissue regeneration in contact with the implant.

Sterilization by exposure to EtO was experimented on chitosan flakes. No degradation was observed, and NMR analysis of EtO-sterilized chitosan flakes did not show any significant chemical change. However, such a sterilization technique cannot be used to sterilize chitosan solutions or hydrogels.

Exposure to saturated water steam is often used for the sterilization of medical devices. Under humid conditions, this sterilization method seems to be ideal for the preparation of sterilized hydrogel-type materials. However, it has been reported that the high temperature, humidity, and pressure used during steam sterilization can lead to hydrolysis, softening, and degradation of biomedical polymers [10].

RMN analysis after material autoclaving did not show any chemical modifications. This sterilizing treatment did not change the DA of chitosan, regardless of the physical form of the polymer. No particular evolution in molecular weight Mw was observed in the case of the sterilization process of chitosan flakes at 134°C during 4 minutes or at 121°C during 20 minutes (see example in figure 1).

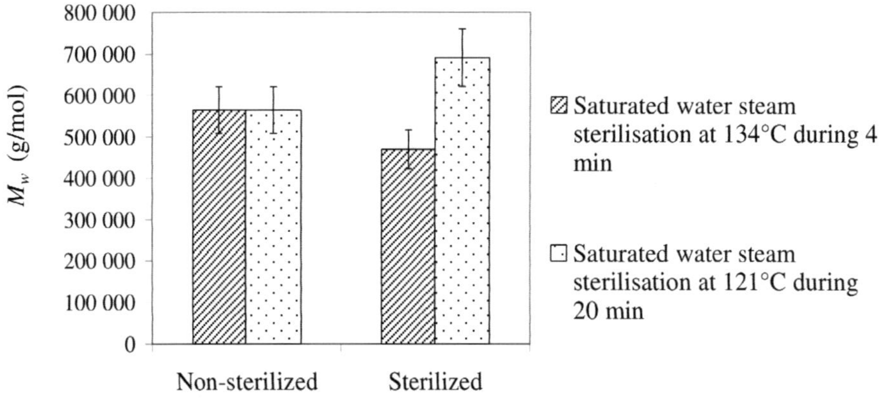

Figure 1. Evolution of Mw for chitosan flake samples (DA = 13.8%) after saturated water steam sterilization.

The sterilization of solutions induced a decrease in molecular weight of chitosan around 60%. The important degradation of chitosan observed in the present study after autoclaving chitosan solutions containing acetic acid is in accordance with No *et al.* that reported autoclaving chitosan solutions in presence of acetic acid as an efficient technique to depolymerize chitosan [11]. This polymer degradation during autoclaving is probably due to the higher molecular mobility in the chitosan solution than in flakes or hydrogels thus facilitating the scission of glycosidic bonds.

Interestingly, exposure of chitosan flakes dispersed in water to saturated water steam at 121°C during 20 minutes did not induce a measurable degradation of the polymer (data not shown).Thus, this method could be considered as an optimal sterilization procedure for the preparation of sterilized chitosan solutions. Nevertheless it implies performing the next steps of material elaboration under sterile conditions.

Except in the case of solutions, exposure to water steam was shown to preserve the macromolecular structure of chitosan. No significant change of M_w, I_p and DA was observed for the different tested samples (hydrogels prepared from hydro-alcoholic or aqueous solutions). The water steam sterilization of the hydrogels after their elaboration is probably the most convenient and cost-saving way to produce sterilized materials.

4. Conclusions

A comparison between several sterilization treatments for different chitosan-based materials has been reported here. It was evidenced that exposure to gamma and beta ionizing radiations should not be advised for any physical form of chitosan because it induces a very important degradation of this polymer.

Exposure of chitosan flakes to EtO was not shown to degrade the polymer but this sterilization technique has limited use for chitosan-based materials to the dry state.

Saturated water steam sterilization was thus shown to be an appropriate method for the treatment of chitosan flakes and for chitosan hydrogels. While exposure of chitosan solutions to saturated water steam leads to a significant depolymerization, autoclaving chitosan flakes dispersed in water seems to be an optimized technique for the preparation of sterilized chitosan solutions.

A more detailed report of this study will be published later.

Acknowledgments

This work was funded by the French National Research Agency with the reference ANR-08-MAPR-0021-01 and by the FUI program MATBIOTEX.

References

[1] Drury J L and Mooney D J 2003 *Biomaterials* **24** 4337-51
[2] Domard A and Domard M 2002 Chitosan: Structure-Properties Relationship and Biomedical Applications (in *Polymeric Biomaterials*; Dumitriu S) ed Marcel Dekker chapter 9.
[3] Boucard N, Viton C, Agay D, Mari E, Roger T, Chancerelle Y and Domard A 2007 *Biomaterials* **28** 3478-88
[4] Montembault A, Tahiri K, Korwin-Zmijowska C, Chevalier X, Corvol M T and Domard A 2006 *Biochimie* **88** 551-64
[5] Ladet S G, Tahiri K, Montembault A S, Domard A J and Corvol M T 2011 *Biomaterials* 2011 **32** 5354-64
[6] Yang Y M, Zhao Y H, Xiao-Hua L, Fei D and Xiao-Song G 2007 *J. Appl. Polym. Sci.* 2007 **104** 1968-72
[7] Montembault A, Viton C and Domard A 2005 *Biomaterials* **26** 933-43
[8] Montembault A, Viton C and Domard A 2005 *Biomacromolecules* **6** 653-62
[9] Hirai A, Odani H and Nakajima A 1991 *Polym. Bull.* **26** 87-94
[10] Zhang Y Z, Bjursten L M, Freij-Larsson C, Kober M and Wesslen B 1996 *Biomaterials* **17** 2265-72
[11] No H K, Nah J W and Meyers S P 2003 *J. Appl. Polym. Sci.* **87** 1890-94

IOP Conf. Series: Materials Science and Engineering **31** (2012) 012008 doi:10.1088/1757-899X/31/1/012008

Novel polycarbonate-graphene nanocomposite foams prepared by CO_2 dissolution

G Gedler, M Antunes, V Realinho and J I Velasco[1]
Centre Català del Plàstic. Departament de Ciència dels Materials i Enginyeria Metal·lúrgica. Universitat Politècnica de Catalunya, C/Colom 114. E-08222 Terrassa, Barcelona, Spain.

[1] E-mail: jose.ignacio.velasco@upc.edu

Abstract. Polycarbonate foams reinforced with 0,5 wt% of graphene were obtained by firstly melt-mixing the polycarbonate and graphene in an internal mixer, compression-moulding the melt-compounded grinded material and lastly dissolving CO_2 inside a high pressure vessel. The CO_2 desorption behaviour in the unfilled polycarbonate and nanocomposite was studied in terms of the CO_2 saturation concentration and desorption diffusion coefficient, with the graphene-filled nanocomposite displaying a higher CO_2 loss rate when compared to the neat polycarbonate. The cellular structure of the foams was found to be highly dependent on the saturation/foaming temperature, with smaller cell sizes being obtained with decreasing the temperature. Another parameter that had an important influence was the residual pressure, with higher residual pressure values resulting in foams with more uniform and regular cells.

1. Introduction

Polymeric foams are currently used in industrial applications where lightness is a key factor [1]. The final properties of these foams depend on their cellular structure, which includes parameters such as the average cell size and size distribution, cell volume fraction and cell arrangement within the matrix [2]. Nevertheless, the use of polymer foams is somewhat limited due to the inherent reduction of their mechanical properties with foaming when compared to the solid base material. These foams are known for displaying better specific properties when cell sizes are reduced to a micrometer scale, hence having the potential to significantly alter the way plastics are employed in a wide variety of applications [3]. Nowadays the trend lies in improving the specific mechanical response of these so-called microcellular foams and adding functionalities, such as thermal or electrical conduction, considering both the material's composition and cellular structure control [4]. The most common microcellular foaming process uses a physical blowing agent that creates an evenly distributed micrometric-sized closed cell structure, which significantly improves the mechanical properties compared to more heterogeneous or open cell structures [5]. Supercritical carbon dioxide (sc-CO_2) is one of the most favorable ones due to its combination of chemical inertness, non-flammability and mild supercritical conditions (T_c = 31 °C, P_c = 7,38 MPa) [6], also being environmentally benign [7]. One of the advantages of using physical blowing agents is that cell structure may be controlled through the processing temperature and pressure. Nevertheless, in order to optimize foam properties, an overall understanding of the gas diffusion and the nucleation and growth mechanisms is required [8].

In order to create cells with uniform sizes, all cells must nucleate simultaneously after rapid depressurization. A common approach to this problem considers the use of fillers that act as heterogeneous nucleating agents [9]. In addition, and depending on their characteristics, these fillers may also act as nanoreinforcements. Carbon nanotubes and nanofibres have increasingly been used looking not only for mechanical but also for thermal and electrical properties improvements. Graphene nanosheets have currently attracted a great deal of interest due to their high theoretical properties and particular flat morphology [10-11]. Graphene displays high mechanical, thermal and electrical properties suitable for developing conductive nanocomposites, which could be useful in electronic circuits, sensors and actuators [12]. The combination of these properties with fire retardancy could lead to the development of polymer nanocomposites with tailor-made properties [13].

Although graphene-reinforced polymer composites have already been thoroughly studied in the last few years, the knowledge of foams made from these materials is still pretty scarce, mainly due to their multiphase nature, direct result of the combination of a complex developed cellular structure and polymer microstructure [14]. The influence of the foaming conditions has lately been considered, mainly focused in obtaining high performance foams by means of decreasing cell size and incorporating inorganic functional fillers [15].

The purpose of this study was to prepare polycarbonate-graphene foams through a physical sc-CO_2 one-step batch foaming process and to characterize them in terms of the developed cellular structure morphology.

2. Experimental

2.1. Materials and compounding

Polycarbonate (Lexan-123R-PC, supplied by *Sabic*), with a density of 1,2 g/cm^3 and MFI of 17,5 dg/min, measured at 300 °C and 1,2 kg, was melt-mixed with 0,5 wt% of graphene using a *Brabender Plasti-Corder* internal mixer. The graphene used in this study was supplied by XG Sciences, Inc., USA. These xGnP-Grade-M graphene nanoplatelets are 6 to 8 nm thick with a 15 μm average diameter and a density of 2,2 g/cm^3, as reported by the manufacturer.

Firstly, the PC and graphene were physically mixed and slowly introduced in the internal mixer at a temperature of 180 °C using a rotating speed of 30 rpm during 2 min. Then the rotating speed was increased to 60 rpm and 120 rpm for 1 and 3 min, respectively. The material was cooled at room temperature, grinded and compression-moulded at 220 °C and 45 bar in a hot-plate press (IQAP LAP PL-15) to discs with a thickness of 3,5 mm and diameter of 74 mm in 3 steps. In the first step the upper hot plate was used to soften the material at a temperature of 220 °C. The second step consisted in allowing air to escape the material by pulsating the upper plate in an up and down motion. The pressure applied from the upper plate was 45 bar at a temperature of 220 °C during 1,5 min. For the third step the material was continuously compressed for 1 min at 220 °C and 45 bar (see Figure 1 for scheme of the compression-moulding process). Lastly, the mould with the sample still inside was left to cool for 15 min in the cooling station of the press applying a constant pressure of 45 bar.

The resulting compression-moulded solid discs were used to prepare the foams by sc-CO_2 dissolution, while the discs used in the CO_2 desorption measurements were prepared by mechanically reducing the disc diameter to a typical value of 40 mm.

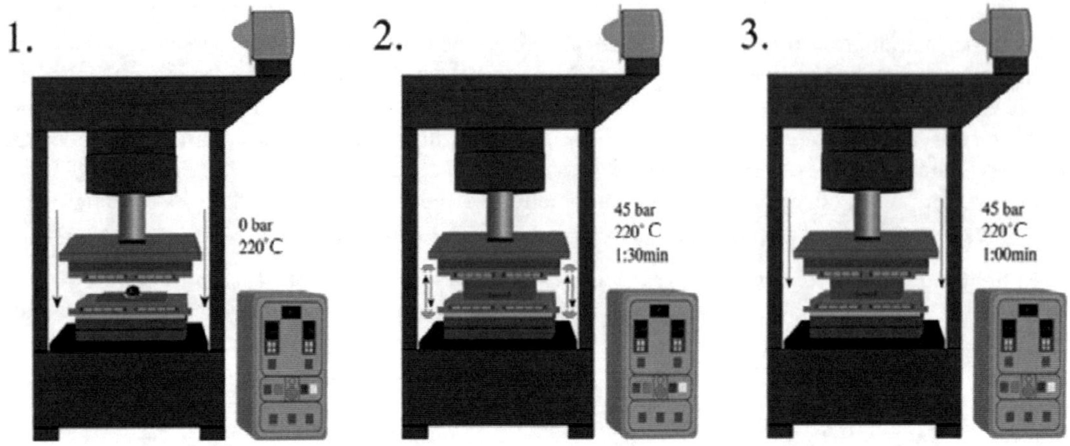

Figure 1. Scheme of the compression-moulding process.

2.2. CO_2 desorption measurements

In order to determine both the solubility and diffusion coefficient of CO_2 in the neat polycarbonate and polycarbonate-graphene nanocomposite, CO_2 desorption measurements were carried out. Samples were introduced in the high pressure vessel and heated up to 210 °C at a CO_2 pressure of 160 bar. After reaching the saturation conditions, the samples were cooled to 40 °C and the CO_2 fully decompressed. Next, the saturated samples were removed from the vessel and quickly transferred to a digital balance (Mettler Toledo PB303 DeltaRange, with a sensitivity of 1 mg) at room temperature and atmospheric pressure, in order to record the CO_2 mass loss as a function of desorption time.

The maximum concentration of CO_2 in the samples after full decompression (M_0) was calculated by extrapolating to zero desorption time using the initial slope method [16]. Assuming one-dimensional diffusion in a plane sheet, the CO_2 desorption diffusion coefficient (D_d) was determined by plotting M_t/M_0 vs. t/l^2, where M_t is the CO_2 concentration at time t and l is the thickness of the sample, according to the following equation [17]:

$$\frac{M_t}{M_0} = 1 - \frac{8}{\pi^2} exp\left(\frac{-D_d t}{l^2}\right) \quad (1)$$

2.3. Foaming process

Foams were prepared by a sc-CO_2 dissolution one-step batch process that consisted in saturating the solid discs with CO_2 inside the high pressure vessel at pressures varying from 120 to 160 bar during time periods between 0 and 120 min. The values of the saturation/foaming temperature varied between 200 and 213 °C, and foaming was done in one-step by applying a sudden pressure drop, keeping residual pressures that ranged between 0 and 20 bar. Figure 2 displays both the CO_2 pressurization and depressurization steps used in foaming.

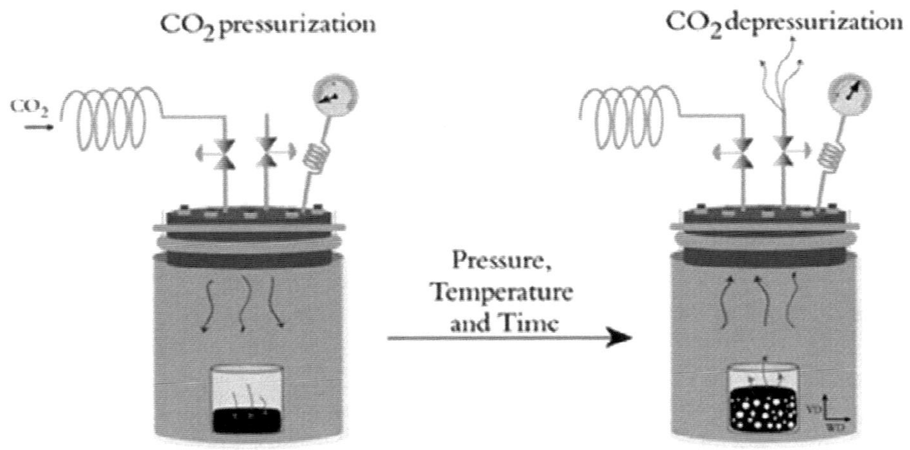

Figure 2. Scheme of the one-step batch foaming process.

2.4. Cellular structure characterization

The cellular structure of the foams was analyzed from scanning electron microscopy images obtained using a JEOL JSM-5610 microscope applying a voltage of 15 kV and a working distance of 30 mm. Samples were previously prepared by fracturing at room temperature and depositing a thin layer of gold onto their surface in argon atmosphere using a BAL-TEC SCD005 Sputter Coater.

The average cell sizes (ϕ) in the vertical (ϕ_{VD}) and width (ϕ_{WD}) foaming directions were measured using the intercept counting method [18]. The cell aspect ratio (*AR*) was determined by dividing the value of the average cell size in the vertical direction by that measured in the horizontal one ($AR = \phi_{VD}/\phi_{WD}$). The cell density (N_f) was calculated using the following equation [19]:

$$N_f = \left(\frac{n}{A}\right)^{3/2}\left(\frac{\rho_s}{\rho_f}\right) \tag{2}$$

Where n is the number of cells per area, A (cm^2), and ρ_s and ρ_f are respectively the solid and foam densities.

3. Results and discussion

3.1. Desorption kinetics

As shown in Figure 3 the maximum concentration of CO_2 dissolved into PC resulted similar for both the unfilled and graphene-filled solid materials. The concentration of CO_2 extrapolated to zero desorption time was respectively 42,6 mg CO_2/g material for the neat PC and 43,0 mg CO_2/g material for the graphene nanocomposite, indicating that the graphene nanoplatelets did not modify the solubility behaviour of CO_2 into PC. Nevertheless, the CO_2 desorption curves showed a very different trend, demonstrating an accuse effect of graphene in the desorption kinetics of CO_2 out of PC, with the graphene nanoplatelets favoring the diffusion of CO_2 out of PC at room temperature. As a result, the calculated desorption diffusion coefficient was $6{,}69\times10^{-11}$ m^2/s for the polycarbonate-graphene nanocomposite, considerably higher than the value of $4{,}45\times10^{-12}$ m^2/s obtained for the neat PC. This value of the CO_2 diffusion coefficient in PC was found to be comparable to others presented in the literature, where D_d was found to be within $2{,}55\times10^{-11}$ and $4{,}60\times10^{-12}$ m^2/s [20] or between $1{,}55\times10^{-12}$ and $6{,}93\times10^{-12}$ m^2/s [21]. This effect of the graphene nanoplatelets in the nanocomposite's structure might be understood as the consequence of the formation of preferential diffusion paths at the surface of the graphene platelets.

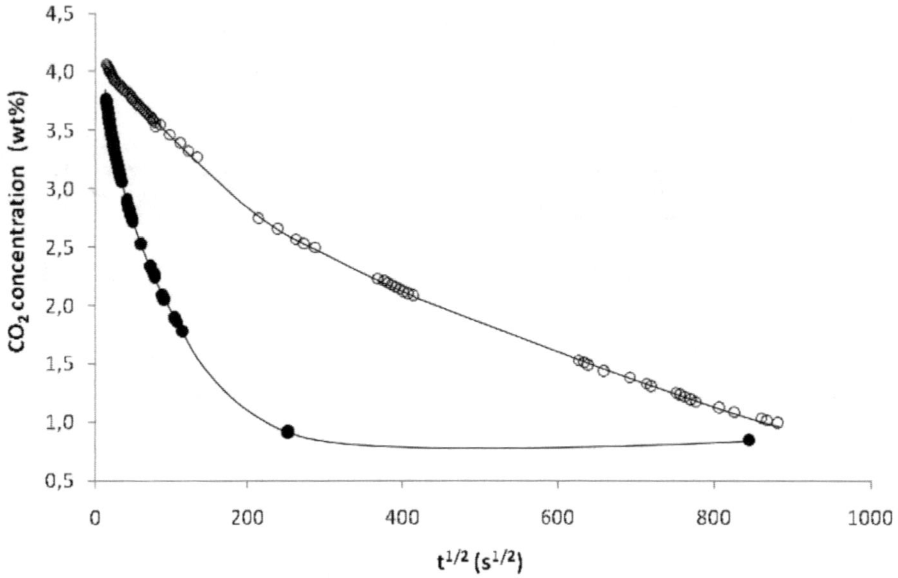

Figure 3. CO_2 desorption curves for the unfilled PC (hollow symbols) and graphene-filled nanocomposite (filled symbols).

3.2. Cellular structure

Table 1 summarizes the values of the foaming parameters for various relative density graphene-filled nanocomposite foams and respective average cell sizes and cell aspect ratios. These values were obtained after analyzing several scanning electron micrographs at different magnifications for each foam (typical SEM micrographs are displayed in Figure 4). As can be seen, a particular cell size value was obtained for each foam at a given saturation/foaming temperature, resulting in a variation in the cell density value from $1{,}04\times10^6$ to $2{,}70\times10^6$ cells/cm^3. As expected, a characteristic homogeneous microcellular structure was obtained due to the CO_2 saturation and sudden depressurization applied during foaming, with both the sudden pressure drop and graphene presence influencing the cell nucleation stage.

Table 1. Foaming process parameters and cellular characterization results of the foamed graphene-filled nanocomposites.

Relative density	T_{sat} [a] (°C)	Time [b] (min)	P_{res} [c] (bar)	ϕ_{VD} (μm)	ϕ_{WD} (μm)	*AR*
0,79	200	20	5	73,0	58,3	1,3
0,59	210	20	10	86,2	97,3	0,9
0,49	205	120	15	92,9	106,2	0,9
0,47	205	40	20	94,1	111,3	0,8
0,38	205	20	10	97,3	103,9	0,9
0,35	213	0	10	146,6	143,8	1,0

[a] Saturation/foaming temperature.
[b] Time at saturation/foaming temperature.
[c] Residual pressure after depressurization.

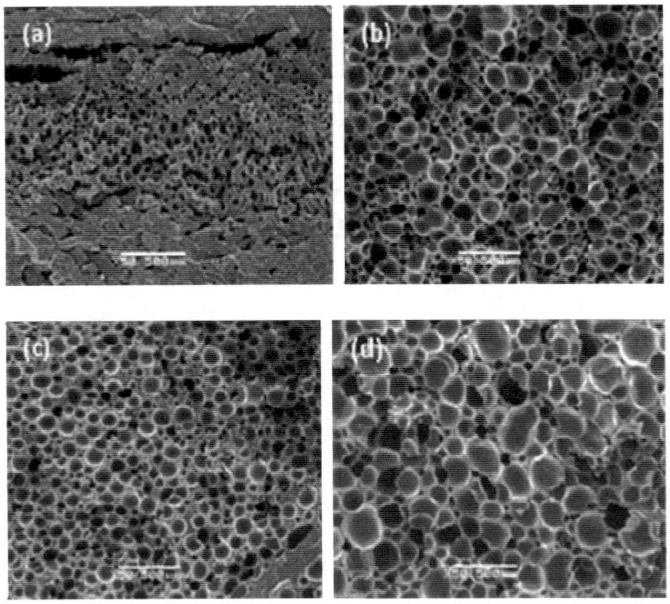

Figure 4. Typical SEM micrographs of the graphene-filled nanocomposite foams obtained at different saturation/foaming temperatures: (a) 200 ºC, (b) 205 ºC, (c) 210 ºC and (d) 213 ºC.

Figure 5 shows how cell size was clearly affected by the saturation/foaming temperature. Generally speaking, cell size increased from 58,3 to 146,6 μm with increasing the saturation/foaming temperature from 200 to 213 ºC, attributed to a higher concentration of CO_2 dissolved in the material. Decreasing the foaming temperature restricted cell growth and resulted in foams with higher relative densities (see Table 1). During the sudden depressurization stage, the temperature of the growing sample decreased quickly, mainly due to the adiabatic expansion of the gas, and in a minor way due to the water cooling system of the vessel. Hence, this behaviour can be explained by a faster rigidization of the softened material foamed at a lower temperature during the depressurization stage, reaching faster the glass transition temperature of the polymer than the samples foamed at higher temperatures.

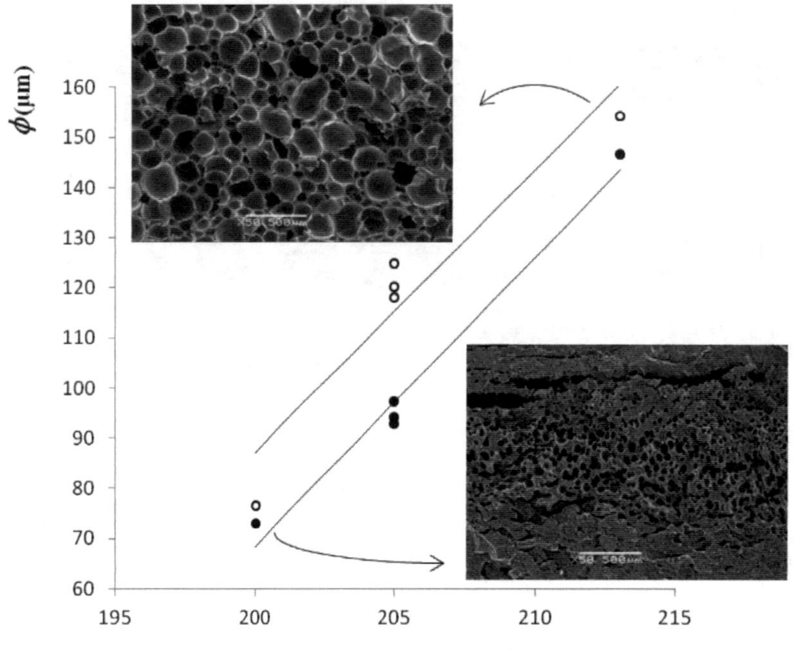

Figure 5. Influence of the saturation/foaming temperature on the average cell sizes of the graphene-filled nanocomposite foams. Hollow symbols: ϕ_{VD}; filled symbols: ϕ_{WD}

The effect of the residual pressure was found to affect the cell aspect ratio (*AR*). Particularly, reducing the residual pressure resulted in an increment of the *AR* (see Figure 6). A residual pressure between 10 and 15 bar was found to be the optimal residual pressure to attain perfectly spherical-like cellular structures (*AR* = 1), while a residual pressure of 5 bar resulted in a deformed cellular structure.

Finally, the observation at high magnifications of the cell walls and struts allowed identifying the presence of graphene nanoplatelets apparently well dispersed in the PC matrix (see Figure 7).

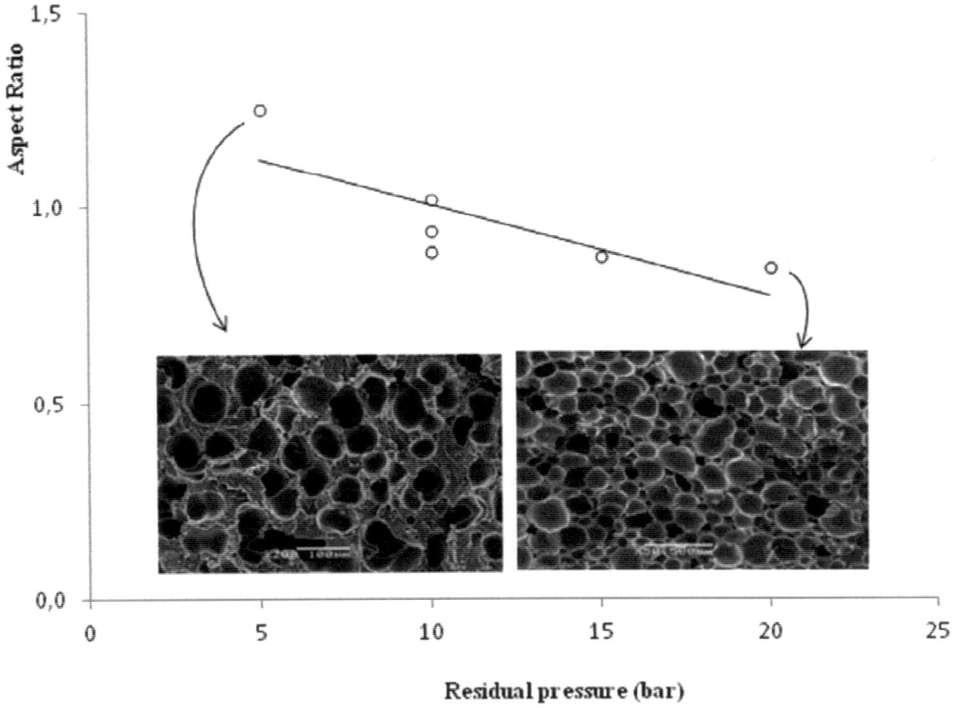

Figure 6. Influence of the residual pressure on the cell aspect ratio of the graphene-filled nanocomposite foams.

Figure 7. Detail of the graphene nanoplatelets.

4. Conclusions

Novel polycarbonate-graphene foams were prepared and characterized in terms of their cellular structure characteristics. The study of the desorption kinetics of CO_2 out of PC and PC with 0,5 wt% graphene showed similar values of the maximum CO_2 concentration dissolved into the material. The graphene-reinforced nanocomposite displayed a higher CO_2 loss rate with a higher diffusion coefficient, related to a preferential diffusion path created at the surface of the graphene nanoplatelets, enabling a faster CO_2 diffusion.

The morphology of foams was found to be highly dependent on the saturation/foaming temperature, with foams displaying smaller cell sizes with decreasing this temperature. This was attributed to a fast cooling of the polymer, which stopped foam growth once the glass transition temperature was reached. Another foaming parameter that resulted relevant in the cellular structure was the residual pressure, which promoted the formation of cells with a uniform and regular shape until stabilization by cooling.

Acknowledgments
The authors would like to acknowledge MICINN (Government of Spain) for the financial support of project MAT2010-15565.

5. References

[1] Landrock A 1995 *Handbook of Plastic Foams* (New Jersey: Noyes) p 222
[2] Yun M and Lee W 2008 *Compos. Sci. Technol.* **68** 202
[3] Kumar V and Weller J 1994 *J. Eng. Ind.* **116** 413
[4] Antunes M, Mudarra M and Velasco J I 2011 *Carbon* **49** 708
[5] Lee L, Zeng C, Cao X, Han X, Shen J and Xu G 2005 *Compos. Sci. Technol.* **65** 2344
[6] Cooper A 2000 *J. Mater. Chem.* **10** 207
[7] Goren K, Chen L, Schadler L and Ozisik R 2010 *J. Supercritical Fluids* **51** 420
[8] Yun M and Lee W 2008 *Compos. Sci. Technol.* **68** 202
[9] Wool R P 2005 *Composites and Foams from Plant Oil-Based Resins* (Bio-Based Polymers and Composites, London, Elsevier Science) p 114
[10] Geim A and Novoselov K 2007 *Nature Materials* **6** 183
[11] Wang F, Zhang Y, Tian C, Girit C, Zetti A, Crommie M and Shen R 2008 *Science* **320** 206
[12] Kuilla T, Bhadra S, Yao D and Kim N 2010 *Prog. Polym Sci.* **35** 1350
[13] Sengupta R, Bhattacharya M, Bandyopadhyay S and Bhowmick A 2011 *Prog. Polym. Sci.* **36** 638
[14] Beechem T and Lafdi K 2006 *Carbon* **44** 1548
[15] Antunes M, Realinho V and Velasco J I 2010 *J. Nanomat.* Article ID 306384 1
[16] Nawaby A and Zhang Z 2005 *Solubility and Diffusivity in Thermoplastic Foam Processing* (Thermoplastic foam processing: Principles and development, London, CRC Press) pp 1-42
[17] Crank J 1956 *The Mathematics of Diffusion* (London: Oxford University Press)
[18] Sims G and Khunniteekool C 1994 *Cell. Polym.* **13** 137
[19] Sorrentino L, Di Maio E and Iannace S 2010 *J. Appl. Polym. Sci.* **116** 27
[20] Tang M, Du T and Chen Y 2004 *J. Supercritical Fluids* **28** 207
[21] Tang M, Huang WH and Chen Y 2007 2007 *J. Chin. Inst. Chem. Eng.* **38** 419

IOP Conf. Series: Materials Science and Engineering **31** (2012) 012009 doi:10.1088/1757-899X/31/1/012009

Microstructure design of UHMWPE-based materials: blending with graphite nanoplatelets.

A Delgado[1,2], F Addiego[1,4], S Ahzi[2], S Patlazhan[3], V Toniazzo[1], D Ruch[1].

[1] Advanced Materials and Structures department, Public Research Centre Henri Tudor, 66 Rue de Luxembourg, L-4221 Esch-sur-Alzette, Luxembourg.
[2] Institut de Mécanique des Fluides et des Solides, Université de Strasbourg, 2 Rue Boussingault, F-67000 Strasbourg, France.
[3] Semenov Institute of Chemical Physics, Russian Academy of Science, Kosygina Street 4, 117977 Moscow, Russian Federation

frederic.addiego@tudor.lu

Abstract. The main scope of this work was to develop a processing method to homogeneously distribute graphite nanoplatelets (GNP) within an ultra-high molecular weight polyethylene (UHMWPE) matrix. After this step, we estimated the toughness of the new nanocomposite material. Combining a sonication step, a micro-extrusion step and a hot-pressing step, we did not reach an optimal distribution state of the nanofillers. Nevertheless, the toughness of the nanocomposites evaluated by Charpy tester was higher than that of the reference UHMWPE (only processed by hot-pressing). We also found that our new processing procedure applied to neat UHMWPE leads to the higher toughness than that of the nanocomposite. Micro-extrusion appears as a promising processing tool for neat UHMWPE.

1. Introduction

Ultra-high molecular weight polyethylene (UHMWPE) is a semicrystalline polymer used as bearing material in non medical applications such as machine construction, sport equipment...and medical applications [1,2] such as the total knee [3] or hip arthroplasty [4,5]. In all these applications, the durability of UHMWPE depends on a subtle combination of strength and toughness that must be as high as possible. One strategy to improve these properties for the UHMWPE based materials is to distribute and disperse nanofillers within the polymer matrix. Due to the extremely high viscosity of UHMWPE, the blending of UHMWPE with nanoparticles by means of screw-extruder or blender is not common. The objective of this paper is to develop a new processing strategy to homogeneously distribute and disperse carbon based nanofillers within UHMWPE and, to characterize the microstructural aspects and the toughness of the new materials by scanning electron microscopy (SEM) and Charpy impact tests. We select for this work graphite nanoplatelets (GNPs) as fillers. It is important to notice that UHMWPE/GNP nanocomposites may not be suitable for prosthesis applications due to the potential toxicity of GNP [6,7]. Nevertheless, the here developed methodology could also be applied to the blending with biocompatible fillers.

[4] To whom any correspondence should be addressed.

2. Materials and methods

2.1. Materials

UHMWPE powder was provided by Ticona (Ticona GmbH, Oberhunsen, Germany) under the reference name GUR 1050 (molecular weight of 5×10^6 g/mol). Graphite nano platelets (GNP), Figure 1, were supplied by Angstron Materials (Dayton, USA) under the reference N006-010-N. These platelets are 10 nm thick and have a diameter which is below 10 µm. The toluene solution used for sonication was supplied by Carl Roth (Karlsruhe, Germany) and has a purity of 99.5 wt. %.).

Figure 1. Raw Graphite nanoplatelets (GNP).

2.2. Processing

2.2.1. Sonication in toluene. GNP powder was first dispersed in a bath of toluene, at 80°C by means of an ultrasonic tip. Once the GNP were dispersed (corresponding to 15 min of sonication), UHMWPE powder was added to the mixture. Afterwards, sonication was pursued to homogenize the mixture and to attain a swelling state of UHMWPE. The mixture colour became homogeneous after 30 min of sonication, indicating the end of the sonication procedure. Finally, toluene was partially eliminated by filtration and 12 h of evaporation at 80°C.

2.2.2. Extrusion. We select a micro-extruder to further distribute and disperse GNP within UHMWPE.This process has been performed with an DSM (Geleen, The Netherlands) micro-extruder (reference Xplore 15 mL) at 220°C, with a screw rotation speed of 150 rpm, and with a die of 3 mm in diameter. These experiments were conducted without recirculation of the polymer. The extruded materials were pelletized into 5 mm long cylinders.

2.2.3. Compression moulding. 10 mm thick circular plates with a diameter of 110 mm were moulded with a Carver laboratory press (Carver Inc, Wabash, USA). The procedure consisted of i) a heating step up to 240°C and with a an initial pressure of 2.5 bars during 30 min and an secondary pressure of 20 bars during 15 min. and ii) a cooling step at 20 bars by a room temperature water circulation within the plates .The references of the materials investigated in this study are reported in Table 1.

Table 1. List of studied materials.

	Precursor material	Process/Treatment
PE	UHMWPE powder	Compression moulding
ePE	UHMWPE powder	Extrusion + compression moulding
ePE/G	UHMWPE powder + 0.5 wt.% of GNP	Sonication + extrusion + compression moulding

2.3. *Scanning electron microscopy*

The microstructure of the materials was observed by means of a pressure-controlled scanning electron microscope (SEM) FEI Quanta FEG 200 (FEI Company Europe BV, Eindhoven, The Netherlands). The electron beam was generated at 5 kV, while the images were recorded by means of a large-field detector (LFD) with the chamber water pressure set to 150 Pa. SEM observations were performed on cryo-fractured and Charpy-broken specimens at the center of the fractured surface.

2.4. *Charpy tests*

Charpy impact tests were performed on 15 specimens of each sample; these specimens were 20 mm thick, 10 mm width, and 55 mm long parallelepipeds. Each specimen was transversally notched at its center to get a ligament section of 10 mm^2 (thickness of 5 mm and width of 20 mm). The notches were done by machining to 8 mm in a Charlyrobot milling machine (Charlyrobot, Cruseilles, France) finally, by means of a fresh razor blade, the notch was sharpened; resulting on a thickness of 5 mm. Charpy tests were carried out at room temperature (23°C) by means of a 300 J Charpy hammer (Schenck-Trebel GmbH, Ratingen, Germany) with a span distance of 40 mm.

3. Results and discussion

3.1. *Toughness of the materials*

The Charpy impact results are listed in Table 2. The reference material showed fracture energy of 17.5 J/cm^2. The sonication and extrusion of neat UHMWPE (ePE) have lead to an increased toughness when compared with the reference material (24.5 J/cm^2 vs. 17.5 J/cm^2). When GNP particles are blended with UHMWPE (ePE/G), we note that the impact energy takes values of 19.6 J/cm^2, which are comprised between those of the reference UHMWPE material and the ones for the neat extruded UHMWPE material. We hypothesize that the increase in toughness of UHMWPE is due to i) an increase in consolidation, ii) an increase in toughness of the matrix and subsequently, iii) the nanofiller relationship within the polymer matrix. The increase in toughness of the polymer matrix may be due to the formation of textured domains with oriented chains due to the combination of extrusion and compression moulding. A preliminary wide angle X-ray scattering (WAXS) investigation could validate the second hypothesis. Regarding the two other hypotheses; we have preformed some microstructural observations which are below described.

Table 2. Impact resistances by Charpy test.

Material	**Energy (J/cm^2)**	**Standard deviation (J/cm^2)**
PE	17.5	1.0
ePE	24.5	1.0
ePE/G	19.6	1.5

3.2. Morphology of the materials

An image of the microstructure for the ePE/G nanocomposites is shown on Figure 2. This image can be compared with the same material at low magnification in Figures 7 and 8; in both, we have observed that the particles seem to be located at the surface of the initial UHMWPE powder grains. Scanning the surfaces at high magnification (Figure 2), we found a limited number of GNP agglomerates. From these results we propound that GNP are poorly distributed within the UHMWPE matrix; while their dispersion appears satisfactory. The here developed processing procedure did not enable us to see such a penetration of GNPs within the UHMWPE grains. The origin of this phenomenon could come from a limited swelling of the UHMWPE macromolecules or, from a low internal mobility which does not induces such interpenetration between GNPs and the swelled macromolecules of UHMWPE. The fact that GNPs are not well distributed indicates that the increase in toughness of ePE/G is not mainly due to the presence of these nanofillers.

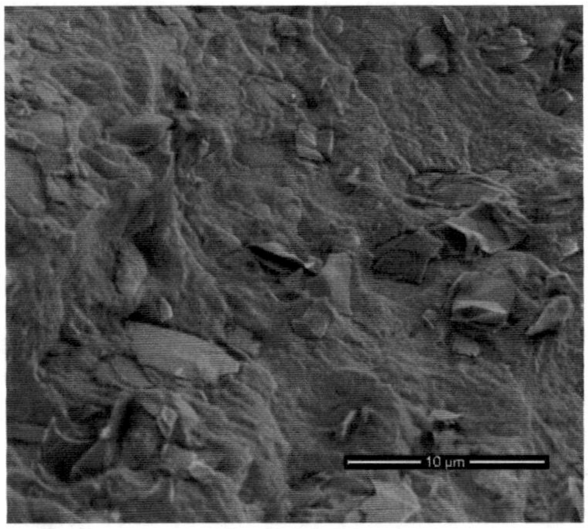

Figure 2. ePE/G morphology.

The consolidation state of the materials can be observed from the SEM micrographs of cryo-fractured materials which are represented in Figures 3, 5 and 7. The material characterized by the higher consolidation aspect is the ePE, since in this material we noticed the smoother fracture surfaces of UHMWPE grains; this suggests the presence of intragranular fracture. Regarding ePE/g and PE, the fractured surfaces seem roughened and to perform a kind of intergranular fracture. The combination of extrusion and compression moulding increases the consolidation state of UHMWPE, which explains the increase in toughness of ePE compared with PE but, the presence of GNPs at the surface of the UHMWPE grains decreases the consolidation state. This phenomenon explains the decrease of toughness when we compare ePE and ePE/g. Regarding the microstructure of the materials fractured by impact (Figure 4, 6, and 8), we have noticed some bands on the PE, ePE and ePE/G surfaces demonstrating such an activation of plasticity during the impact testing for the entire materials. We suppose that the presence of textured domains in ePE and ePE/G would disturb the propagation of cracks and hence, would increase the resistance to fracture of the extruded materials. This phenomenon will be largely investigated in a further paper.

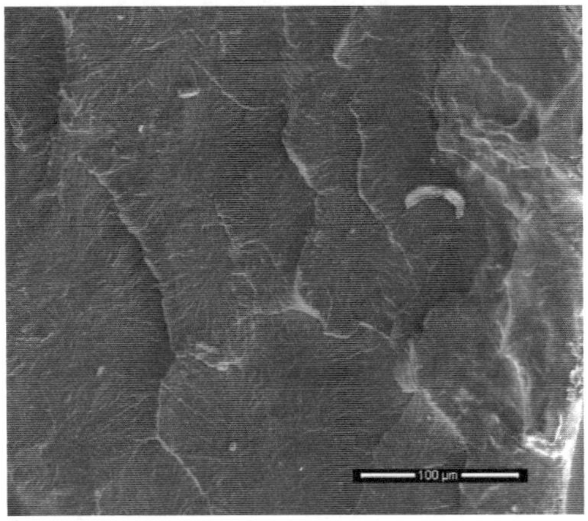

Figure 3. PE cryo-fractured.

Figure 4. PE fractured by impact testing.

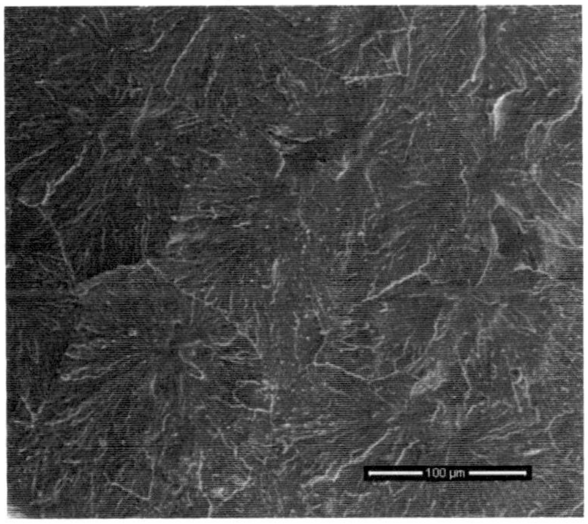

Figure 5. ePE cryo-fractured.

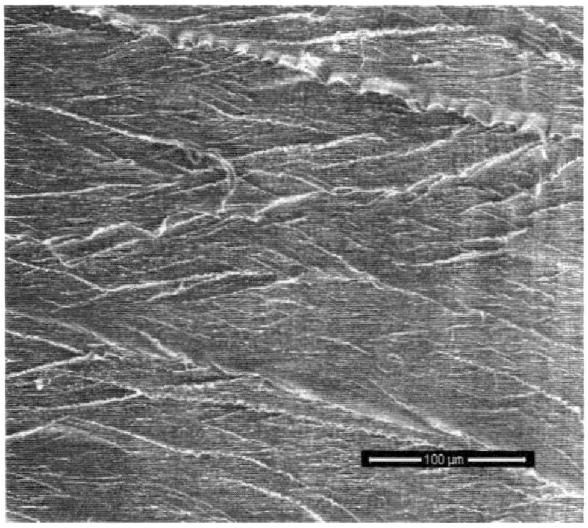

Figure 6. ePE fractured by impact testing.

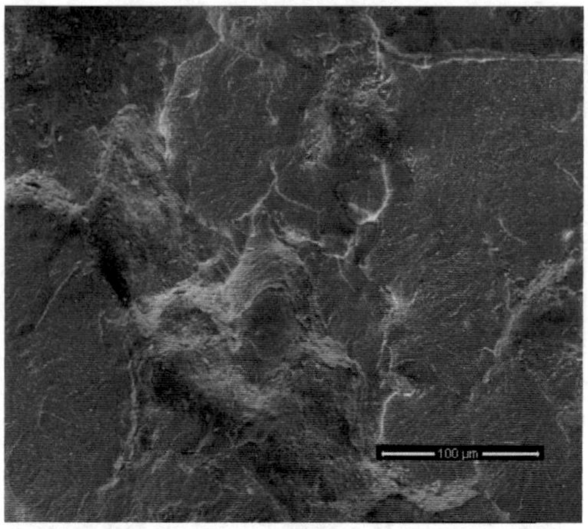

Figure 7. ePE/G cryo-fractured.

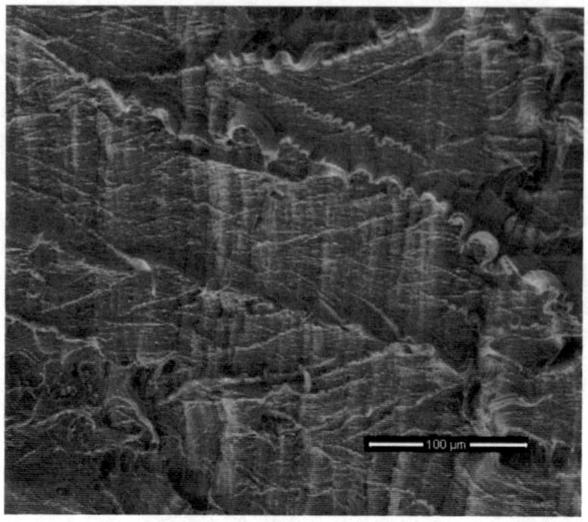

Figure 8. ePE/G fractured by impact testing.

4. Conclusions

The processing procedure developed in this work does not seem to enable homogeneously distribute graphite nanoplatelets within UHMWPE. Nethertheless, the toughness of the UHMWPE/graphite nanoplatelet composite is higher than for the reference UHMWPE. This result could be explained by an increase in the toughness of the UHMWPE matrix due to the presence of textured domains induced by the combination of extrusion and compression moulding processes. The most interesting results are obtained for the neat UHMWPE elaborated by our new processing procedure. Indeed, this material exhibits the highest toughness. The last could be explained by both the presence of textured domains and a high consolidation state (compaction of the materials). The combination of extrusion and compression moulding appears hence as a promising procedure to increase the toughness of UHMWPE, while the use of nano fillers appears to have an undesirable effect in the impact resistance of the material. The generality of these results are currently investigated by a wider range of experimental conditions.

Acknowledgments

Authors wish to acknowledge the financial support from the National Research Fund, Luxembourg. We also would like to thank Ticona Benelux for providing the UHMWPE resin.

5. References

[1] Kurtz, S.M., 2009, A Primer on UHMWPE, in UHMWPE Biomaterials Handbook (Second Edition), *Academic Press: Boston.* p. 1-6.
[2] Bracco, P.; Brunella, V.; Luda, M. P.; Zanetti, M.; Costa, L., 2005, *Polymer*, 46(24): p. 10648-10657.
[3] Rodriguez, J.A., 2008, *The Journal of Arthroplasty*, 23(7, Supplement 1): p. 31-34.
[4] Manley, M.T. and K. Sutton, 2008, *The Journal of Arthroplasty,* 23(7, Supplement 1): p. 47-50.e1.
[5] Kurtz, S. M.; Muratoglu, O. K.; Evans, M.; Edidin, A. A., 1999, *Biomaterials*, 20(18): p. 1659-1688.
[6] Ingham, E. and J. Fisher, 2000, *Proceedings of the Institution of Mechanical Engineers, Part H: Journal of Engineering in Medicine*, 214(1): p. 21-37.
[7] Tipper, J.L., 2009, *4th UHMWPE International Meeting,* Torino, ITALY: UHMWPE Lexicon.

IOP Conf. Series: Materials Science and Engineering **31** (2012) 012010 doi:10.1088/1757-899X/31/1/012010

Martensitic Transformation during Fatigue Testing of an AISI 301LN Stainless Steel

A Mateo[1], G Fargas and A Zapata
CIEFMA, Departament de Ciència dels Materials i Enginyeria Metal·lúrgica
Universitat Politècnica de Catalunya, Avda Diagonal 647, 08028 Barcelona, Spain

E-mail: Antonio.manuel.mateo@upc.edu

Abstract. The plastic deformation accumulated during fatigue testing can induce the transformation of austenite to martensite in metastable austenitic stainless steels. To analyze this issue, a metastable austenitic stainless steel grade AISI 301 LN was studied in two different conditions, i.e. annealed and cold rolled. In the first case, the steel was fully austenitic, whereas cold rolled material had almost 30% of martensite. High cycle fatigue tests at a stress ratio of 0.8 were carried out on flat specimens from both steel conditions. Several characterization techniques, including optical microscopy, X-ray diffraction (XRD) and electron back scattered diffraction (EBSD), were used to detect and quantify the martensite induced by the cyclic deformation.

1. Introduction

Metastable stainless steels are currently used in applications where severe forming operations are required due to their high corrosion resistance combined with versatile mechanical properties [1]. In the annealed state, these steels have a totally austenitic microstructure which confers them an excellent ductility although limited strength. Cold deformation generates strain-induced martensite, in addition to ordinary strain hardening of austenite, resulting in a drastic improvement of yield stress but ductility is consequently decreased [2-6].

Two types of martensite can be induced by plastic deformation: hexagonal-close-packed (ε) and body-center-cubic (α') [7]. Previous studies [2, 3] found that ε-martensite was formed at the beginning of deformation but its volume fraction decreases to almost zero at 20% of plastic strain. On the other hand, α'-martensite increases continuously and at high strain values becomes the only type of martensite present in the steel.

Studies on the fatigue response of metastable stainless steels reported different behaviors depending on the testing conditions [8-12]. The formation of martensite during deformation is known to be harmful in the low cycle fatigue (LCF) regime, i.e. under strain-control, while a small amount of martensite can be beneficial in the high cycle fatigue (HCF) regime, i.e. tests under load-control. Concerning fatigue crack growth (FCG), Chanani and Antolovich [8] concluded that martensite retards crack propagation.

In a previous paper presented by the authors to the last EEIGM International Conference [13], the possibility of improving the fatigue life of an austenitic stainless steel by inducing martensitic transformation via torsion testing was investigated. Continuing with this order of ideas, the present

[1] To whom any correspondence should be addressed

IOP Conf. Series: Materials Science and Engineering **31** (2012) 012010 doi:10.1088/1757-899X/31/1/012010

investigation was undertaken with aim of comparing the HCF performance of a metastable austenitic stainless steel under two different initial conditions: annealed and cold rolled.

2. Material and experimental procedures

2.1. Material

The stainless steel used in this study was a commercial grade AISI 301 LN (corresponding to standard EN 1.4318). It was produced by OCAS NV, Arcelor-Mittal R&D Industry Gent (Belgium). The nominal chemical composition of this grade and also the values obtained from an analysis by arc emission spectrometry are given in Table 1.

Table 1. Chemical composition of AISI 301LN steel (in wt %).

	C	Cr	Ni	Mn	N	Si	Mo	P	S
Nominal	$\leq$ 0,03	16,5-18,5	6-8	$\leq$ 2	0,1-0,2	$\leq$ 1	-	0,045	$\leq$ 0,015
Analyzed	0,015	17,36	7,18	1,68	-	0,55	1,68	0,052	0,006

AISI 301LN sheets of 1.5 mm in thickness were supplied in two different conditions: annealed and cold rolled. In the first case, steel was subjected to cold rolling, annealing, pickling and a final skin passing. In the second one, the process was the same but an additional cold rolling step with 30% of reduction was done.

2.2. Metallography

In metastable steels, mechanical grinding and polishing can induce martensitic transformation on the surface layer. In order to avoid this effect, electro-polishing with 65% nitric acid solution was carried out at 12 V. Following, electro-etching in the same solution at 1,5 V was used to discover the austenite microstructure, whereas martensitic phase was revealed by etching in Beraha's solution [14]. Samples were observed using an Olympus LEXT light optical microscope with confocal laser scanning mode.

2.3. Fatigue testing

In order to fix the fatigue testing parameters, mainly the maximum load to apply, it was necessary to know the monotonic properties of the 301LN steel. Therefore, specimens with the load axis parallel the rolling directions were cut from the sheets by laser. Their main dimensions were: gauge length of 28 mm, 6 mm wide and 1.5 mm thick. Tensile tests were performed by using an Instron universal testing machine at a crosshead rate of 3 mm/min. Table 2 gives the results for the two steel conditions. As expected, the cold rolled steel exhibits higher strength, particularly yield stress, but lower ductility than the annealed material.

Table 2. Results of tensile tests at room temperature.

Steel Grade	Product Condition	Yield Stress, σ_{ys} (MPa)	Ultimate tensile strength, σ_{uts} (MPa)	Elongation to fracture (%)
301LN	Annealed (AN)	360	902	40
301LN	Cold Rolled (CR)	926	1113	24

Considering the tensile properties, a fatigue testing procedure was stated. It consisted on starting the tests applying a maximum load (σ_{max}) of 80% of the ultimate tensile strength of the corresponding steel condition and afterwards, if the specimen has been able to reach $2x10^6$ cycles without fail, σ_{max} is increased 5%, and so on until fracture. The fatigue limit is understood as the stress limit at which the material can be cycled for a sufficiently high number of cycles (usually more than 10^6 cycles) without fracture. In the present case, fatigue limit was considered as the highest stress level at which a specimen did not fail after $2x10^6$ cycles.

Flat fatigue specimens with hour-glass shape (3.8 mm wide and 1.5 mm thick) were laser machined from the steel sheets. Their load axis coincided with the rolling direction, as in the case of tensile specimens. Tests were conducted under load control in a resonant testing machine Rumul Mikroton, working at frequencies around 150 Hz. The imposed stress ratio (R= $\sigma_{min}/\sigma_{max}$) was 0.8. This value was chosen based on previous investigations carried out by the authors that demonstrated a more significant effect of strain induced martensite for elevated stress ratios [13].

After determining the fatigue limit for each steel condition, values under those limits were chosen in order to perform tests without reaching fracture to analyze the martensite formation during cyclic loading. These tests were stopped after 10^6 and $3x10^6$ cycles and subsequently the specimens were examined by optical microscopy, XRD and EBSD techniques.

2.4. Diffraction methods

2.4.1. X-ray diffraction. X-ray diffraction measurements were performed by using Copper radiation on a Brucker D8 Advance equipment. Determination of martensite content was carried out by the method corresponding to reference intensity ratio (RIR) (ASTM E975-03 [15]). This method allows determining the mass fractions of austenite and martensite by using Equation (1).

$$\frac{X_{\alpha'}}{X_\gamma} = \frac{RIR_\gamma}{RIR_{\alpha'}} \times \frac{I_{\alpha', observed}}{I_{\gamma, observed}} \times \frac{I_{\gamma, reference}}{I_{\alpha', reference}} \quad (1)$$

Where "$X_{\alpha'}$" and "X_γ" are the mass fractions of α'-martensite and γ-austenite, respectively; "RIR_γ" and "$RIR_{\alpha'}$" are their respective reference intensity ratios; "$I_{observed}$" and "$I_{reference}$" are the observed and the reference intensities [16].

2.4.2. Electron back scattered diffraction. EBSD scans were performed in a JSM-7001F field emission scanning electron microscope (FESEM) equipped with Channel 5 system (HKL Technology), operating at 20 kV with samples tilted 70 degrees. Before EBSD observations, samples were mechanically and electrolytically polished. Also, after fatigue testing all specimens were again electrolytically polished to reduce its surface roughness which hindered the analysis.

3. Results

The fatigue limits obtained following the testing procedure described in subsection 2.3 are given in Table 3. Both maximum load (σ_{max}) and stress amplitude ($\Delta\sigma/2$) are indicated. Fatigue sensitivity, defined as the ratio between $\sigma_{f,max}$ and σ_{uts}, is higher than 1 for both material conditions. This means that the steel is able to resist more than one million of cycles despite values higher than its monotonic resistance are reached on each fatigue cycle.

Table 3. **Fatigue limits of annealed and cold rolled 301LN steel.**

Steel	Fatigue Limit, $\sigma_{f,max}$ (MPa)	Fatigue Limit, $\Delta\sigma/2$ (MPa)	Fatigue Sensitivity, $\sigma_{max}/\sigma_{uts}$
AN	969	97	1.07
CR	1197	120	1.08

Considering the fatigue limits reported in Table 3, the maximum stress values chosen in order to perform tests without reaching fracture were σ_{max} = 720 MPa for annealed steel and 1100 MPa for cold rolled material.

A quite homogeneous microstructure composed of equiaxial austenitic grains (average grain size of 7 μm) with typical annealing twins is observed in Figure 1(a), corresponding to annealed 301LN. Figure 1(c) belongs to the same condition but etched with Beraha's reagent trying to reveal the presence of martensite, nevertheless almost nothing of this phase is appreciated.

Figure 1(b) shows that after cold rolling the austenitic microstructure had almost not changed. On the other hand, selective etching (Figure 1(d)) revealed a significant proportion of martensite. However, classical light microscope did not allow distinguishing between martensite lathes, slip lines and deformation twins, and therefore it was not possible to quantify the martensite volume fraction by image analysis.

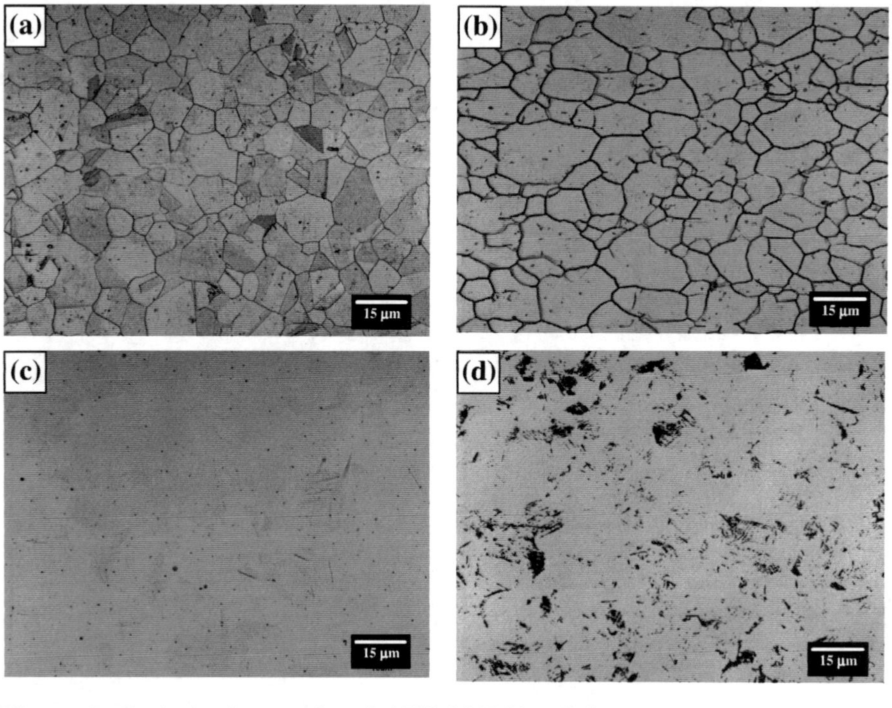

Figure 1. Optical micrographs of AISI 301LN stainless steel in: (a) annealed condition; (b) cold rolled condition; (c) annealed condition with Beraha's etch to reveal martensite; (d) cold rolled condition with Beraha's etch to reveal martensite.

Figure 2 (a) and (b) compare the X-ray spectra of AISI 301LN stainless steel in both initial conditions and also after fatigue testing during 10^6 cycles. Before cyclic loading, and in agreement with the optical microscopy observations, the annealed steel exhibited a totally austenitic spectrum each spectrum, without peaks associated neither with α'-martensite nor with ε-martensite. The cold rolling condition exhibits, together with austenite peaks, α´-martensite ones corresponding to the phase transformation induced by this forming operation. The proportion of martensite was quantified as 28%.

After fatigue testing, in the spectra corresponding to annealed steel new peaks appeared that were indexed as α´-martensite. Calculations of this phase content gave 44% after 10^6 cycles and this value increased up to 50% after $3x10^6$ cycles. Concurrently, Figure 2 (b) shows an increase of martensite peaks intensities (mainly in {211} α' peak) for cold rolled steel after being subjected to fatigue. The initial value of 28% augmented up to 57% and this martensite content remained constant for 1 and 3 millions of fatigue cycles.

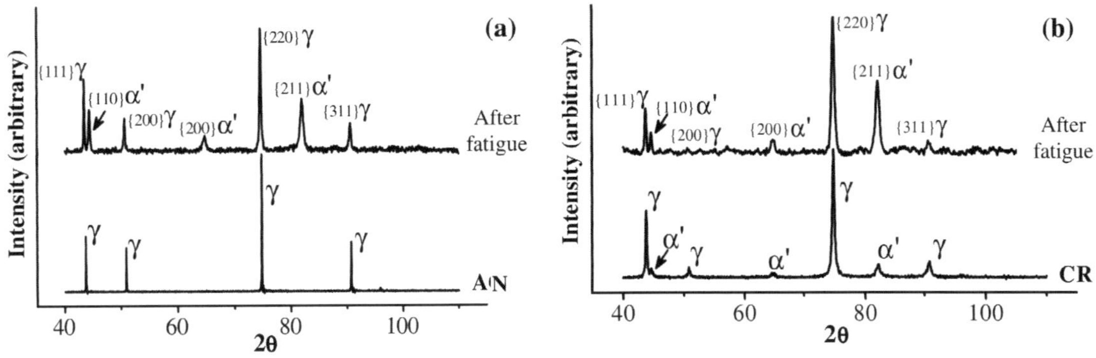

Figure 2. XRD spectra of AISI 301LN stainless steel in (a) annealed and (b) cold rolled conditions, before and after fatigue testing during 10^6 cycles.

Figure 3a shows the EBSD maps of α'-martensite phase corresponding to the annealed steel after 10^6 cycles. In contrast to optical microscopy, EBSD allows to characterize the present phases. Accordingly, grey areas are γ phase, red lines indicate the twins and the coloured zones are martensite. Figure 3b presents magnifications of selected zones of Figure 3a, to evidence that α'-martensite preferentially nucleates at the austenitic grain boundaries or in connexion with the twins, but also appears covering entire grains. The orientation of martensite variants is in accordance with the colours legend in Figure 3b4. The predominant texture corresponds to direction <111> by the great number of blue and violet grains.

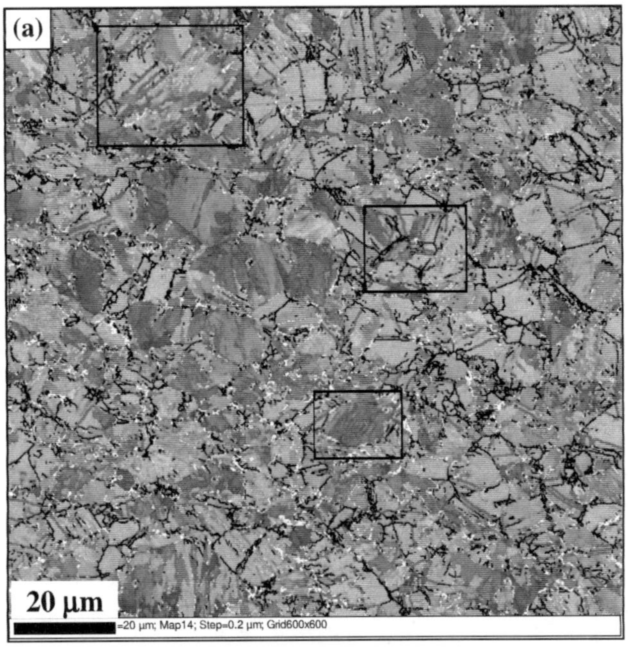

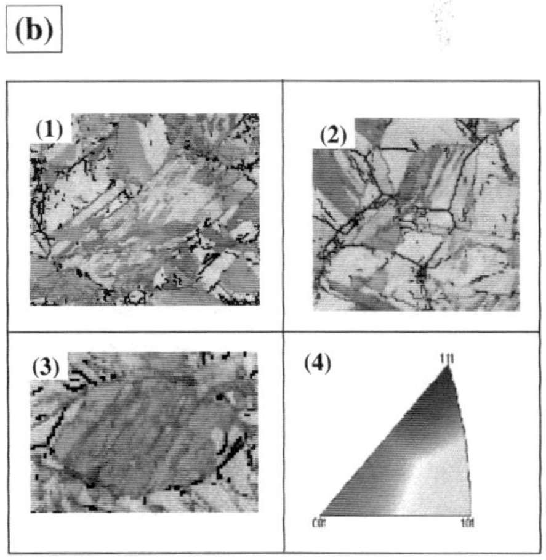

Figure 3. EBSD orientation map of (a) □α'-martensite phase in the annealed condition after fatigue tested at 1×10^6 cycles. (b1-3) Grains zoom and (b4) color legend.

One advantage of EBSD technique is that it gives local information. The local misorientation component indicates small orientation changes on the map, highlighting the region of higher

IOP Conf. Series: Materials Science and Engineering **31** (2012) 012010 doi:10.1088/1757-899X/31/1/012010

deformation. The component calculates the average misorientation between every pixel and its surrounding pixels, and assigns the mean value to that pixel. Figure 4a shows a local misorientation map of austenite in the annealed steel. The homogeneous shade indicates a small misorientation between the different grains and fully austenitic microstructure. As the steel is subjected to deformation, either on cold rolling or during fatigue testing, austenite grains rotate and misorientation angle increases, as plotted in Figure 4b.

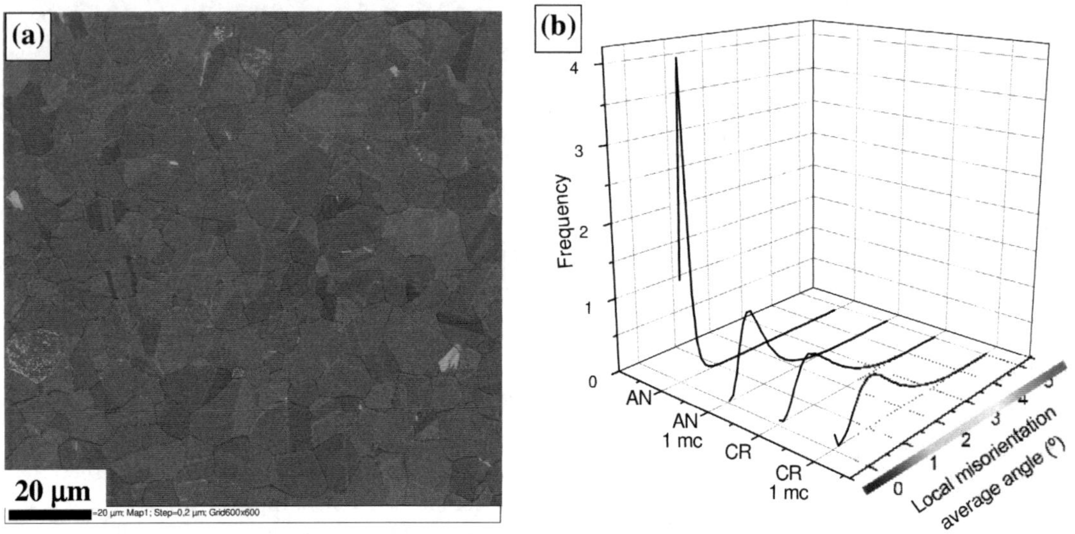

Figure 4. (a) Misorientation map of γ-austenite phase in the annealed condition and (b) histograms of γ-austenite phase for each steel.

4. Conclusions

Specimens of metastable austenitic stainless steel AISI 301 LN in two different initial conditions, i.e. annealed and cold rolled, were subjected to high cycle fatigue tests at a stress ratio of 0.8 in order to establish the fatigue limits. In both cases, the determined fatigue limits (in terms of σ_{max}) were higher than the respective ultimate tensile strengths. This surprising fact is due to the gradual microstructural evolution during fatigue testing due to the formation of strain-induced martensite from the original austenite. XRD analysis revealed the presence of only one type of martensite, i.e. α'-martensite, whereas EBSD images showed that preferential sites for α' nucleation were twins and grain boundaries.

In addition, some tests were performed at cyclic conditions in the stress range below the fatigue limit. Annealed steel was initially fully austenitic, but its capability to be plastically deformed locally, even at stresses well below the fatigue limit, activated the austenite to martensite transformation. The amount of martensite measured by XRD arrived to 50%.

Cold rolled steel had 28% of α'-martensite induced by the rolling process. After cyclic loading below the fatigue limit, it seems that the martensite content reached a saturation level of 57%.

Further investigations are in progress in order to establish if there is a critical amount of martensite to enhance fatigue resistance.

5. References

[1] Lamb S 2001 *CASTI Publishing Inc.* (Canada-ASM International) 473p
[2] Mangonon P L and Thomas G 1970 *Metall. Trans.* **1** 1587-94
[3] Seetharaman V and Krishnan 1981 R *J. Mater. Sc.* **16** 523
[4] Tavares S S M, Fruchart D and Miraglia S A 2000 *J. All. Comp.* **307** 311-17.

[5] Hedström P 2007 *Deformation induced martensitic transformation of metastable stainless steel AISI 301* Licentiate Thesis (Luleå-Sweden) 89p
[6] Mertinger V Nagy E, Tranta F and Sólyom J 2008 *Mater. Sc. Eng. A* **A481-A482** 718-22
[7] Gey N, Petit B and Humbert M 2005 *Metall. Mater. Trans. A* **36A** 3291-99
[8] Chanani G R, Antolovich S D and Gerberich W W 1972 *Metall. Trans.* **3** 2661-72
[9] Baudry G and Pineau A 1977 *Mater. Sc. Eng.* **28** 229-42
[10] Bathias C and Pelloux R M 1973 *Metall. Mater. Trans. B* **4** 1265-73
[11] Stolarz J, Baffie N and Magnin T 2001 *Mater. Sc. Eng. A* **A319-A321** 521-6
[12] Topic M, Tait R B and Allen C 2007 *Internal. J. Fatig.* **29** 656-65
[13] Fargas G, Zapata A, Anglada M and Mateo A 2009 *IOP Conf. Ser.: Mater. Sci. Eng.* **5** 012008
[14] Vander Voort G F 1999 *Metallography: principles and practice / George F. Vander Voort* (Ohio-ASM International) p 557
[15] ASTM E 975-03. *Standard Practice for X-Ray Determination of Retained Austenite in Steel with Near Random Crystallographic Orientation.* ASTM Internacional, 2003.
[16] Beese A M 2008 *Quantification of Phase Transformation in Stainless Steel 301LN Sheets* M. Sc. Thesis (Massachusetts-United State) 107p

Acknowlegment
This work was carried out in the frame of the research project MAT09-14461 (Ministerio de Ciencia e Innovación). Authors acknowledge the support from the *Agència de Gestió d'Ajuts Universitaris I de Recerca* (AGAUR). A.Z. thanks FI-DGR for her fellowship. OCAS NV, ArcelorMittal R&D Industry Gent (Belgium) is recognized for supplying the AISI 301LN sheets.

IOP Conf. Series: Materials Science and Engineering **31** (2012) 012011 doi:10.1088/1757-899X/31/1/012011

Weldability charts for constructional steels

J C Ion[1] and M F Ashby[2]

[1] Division of Materials Science, Luleå University of Technology, SE-97187 Luleå, Sweden
[2] Department of Engineering, University of Cambridge, Cambridge CB2 1PZ, UK

E-mail: john.ion@ltu.se

Abstract. The weldability of materials is still a poorly understood concept; a quantitative assessment remains elusive. The variables associated with welding are reduced here into two groups - processing parameters and material properties - from which two characteristic indices are defined and used as the basis of weldability charts. For the case of constructional steels, a carbon equivalent characterises both heat affected zone hardenability and the maximum hardness developed after solid state phase transformations. The welding process is characterised by its energy input. A mathematical model is used to establish relationships between the indices, which are displayed on charts as contours of microstructure and hardness.

1. Introduction

Despite extensive research, the weldability of steels is still a poorly understood concept. It may be defined in terms of various criteria: hardenability, sensitivity to cracking, limits on distortion etc. Even within the context of the hardenability of steels, different interpretations are applied: a maximum acceptable hardness, or a microstructure containing a given amount of martensite. Unlike the mechanical behaviour of materials, which follows well established rules that can be used to define material performance indices [1], a numerical measure of weldability remains elusive. Attempts have been made to model the effect of weld energy input on the properties of the heat affected zone (HAZ) of carbon steels, which led to the development of *welding diagrams* [2-4]. Such treatment enables weld properties to be displayed graphically in terms of the principal process variables.

Here we define weldability in terms of two quantities that characterise i) the material being welded and ii) the welding process. The *material index* describes both the hardenability of the steel and the hardness of solid state transformation products, and is based on composition through the use of a simple carbon equivalent formula. The *process index* characterises the property of the welding process most relevant to weldability - its energy input. Figure 1 shows a chart constructed using these indices, in which welding processes are displayed based on their energy, and steels are indicated according to typical ranges of carbon equivalent. Below we describe how the indices can be used with analytical models of thermal cycles induced in the HAZ adjacent to the fusion line (where hardness reaches a maximum) and in empirical descriptions of the subsequent phase transformations. Model-based relationships between the indices inform the features of the *weldability charts* that are constructed.

[1] To whom correspondence should be addressed

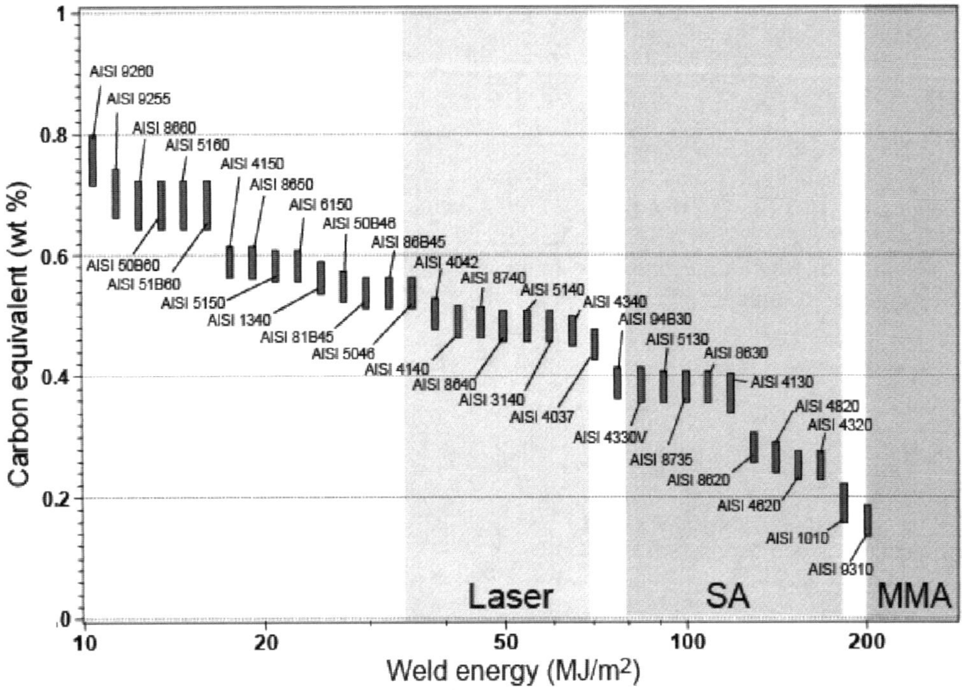

Figure 1. Empirical chart showing carbon equivalent ranges for various steels (bars) and typical ranges of weld energy (energy flux) for various processes (boxes). SA = submerged arc, MMA = manual metal arc. Note that weld energy is measured in MJm^{-2} and MJm^{-1} for thin (full penetration) and thick (partial penetration) welding, respectively. There is no relation in this chart between the axes.

2. Modelling of weldability

2.1. Weld thermal cycles

The thermal cycle $T_{(r,t)}$ at a point in a fully penetrating weld HAZ (e.g. a laser weld) is [5]:

$$T_{(r,t)} - T_0 = \frac{Aq}{vd}\frac{1}{(4\pi\lambda\rho ct)^{1/2}}\exp - \frac{r^2}{4at} \tag{1}$$

where T_0 is initial (or preheat) temperature, A is fraction of incident energy absorbed by the weld, q is weld power, v is welding speed, d is plate thickness, r is lateral distance from the energy source, and for steels λ is thermal conductivity (41 $Wm^{-1}K$), ρ is density (7800 kgm^{-3}), c is specific heat capacity (420 $Jkg^{-1}K$), and a is thermal diffusivity (12.52 x 10^{-6} m^2s^{-1}) [2].

Provided that the peak temperature lies above about 1000°C (i.e. the point of interest lies in the HAZ close to the weld fusion line), equation (1) may be differentiated and solved to obtain the cooling time between 800 and 500°C in the HAZ, Δt, for a two dimensional (thin plate) state of heat flow:

$$\Delta t = \left(\frac{Aq}{vd}\right)^2 \frac{1}{4\pi\lambda\rho c\theta_1^{\,2}} \tag{2}$$

where

$$\frac{1}{\theta_1^{\;2}} = \frac{1}{(773-T_0)^2} - \frac{1}{(1073-T_0)^2}. \tag{3}$$

Corresponding equations may be derived in a similar manner to describe heat flow in the HAZ produced by a surface point energy source (e.g. manual metal arc, metal inert gas, tungsten inert gas or submerged arc welding), in which r becomes the radial distance from the energy source in a three-dimensional (thick plate) temperature field:

$$T_{(r,t)} - T_0 = \frac{Aq}{v}\frac{1}{2\pi\lambda t}\exp - \frac{r^2}{4at} \tag{4}$$

$$\Delta t = \left(\frac{Aq}{v}\right)\frac{1}{2\pi\lambda\theta_2} \tag{5}$$

where

$$\frac{1}{\theta_2} = \frac{1}{(773-T_0)} - \frac{1}{(1073-T_0)}. \tag{6}$$

Equations (2 and 5) indicate that the quantities $Aq/(vd)$ (thin plate welding) and Aq/v (thick plate welding), which are measures of the weld energy input, characterise the cooling rate in the HAZ within the austenite transformation range. These are used as the process index.

2.2. *Phase transformations in the HAZ*

Diffusion-controlled transformation of austenite to a transformation product i is modelled using the form of the Johnson-Mehl equation relevant to nucleation on grain edges [6]:

$$V_i = 1 - \exp\text{-}\{0.69.(\Delta t/\Delta t_i^{50})^2\} \tag{7}$$

where V_i is the volume fraction of i formed after a time Δt, and Δt_i^{50} is the time required for half the transformation to occur. Figure 2 illustrates microstructural development as a function of cooling time in steels and shows the times required for a given amount of transformation.

The transformation of bainite is thus described by:

$$V_b = 1 - \exp\text{-}\{0.69.(\Delta t/\Delta t_b^{50})^2\} \tag{8}$$

where V_b is the volume fraction of bainite formed after a time Δt, and Δt_b^{50} is the time required for half the transformation to occur. The term Δt_b^{50} is not always clearly defined in published data [7], and so is expressed as a function of the time resulting in a bainite-free microstructure, Δt_b^0, and the time for a ferrite-free microstructure, Δt_f^0; quantities that can be obtained easily from continuous-cooling-transformation (CCT) diagrams:

$$\Delta t_b^{50} = \exp(\ln(\Delta t_b^0.\Delta t_f^0)^2/2). \tag{9}$$

Any remaining austenite is assumed to transform to martensite. Microstructural transformation is then characterised by:

$$V_m = 1 - \exp\text{-}\{0.69.(\Delta t/\Delta t_m^{50})^2\} \tag{10}$$

where V_m is the volume fraction of martensite formed for cooling with a characteristic time Δt, and Δt_m^{50} is the characteristic cooling time that results in a microstructure containing 50% martensite.

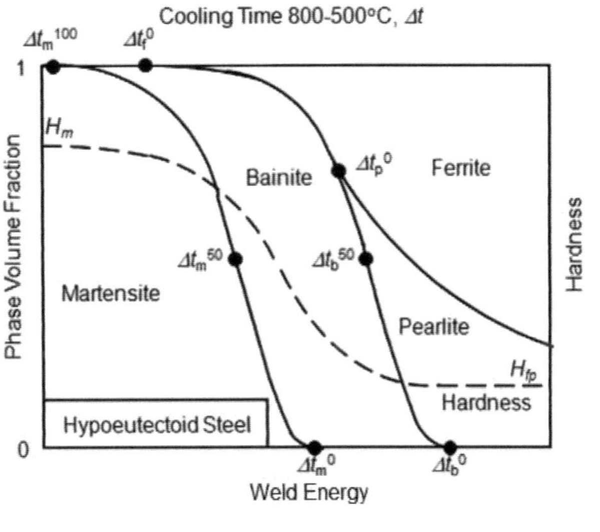

Figure 2. Chart showing model-based variation of microstructure and hardness with weld energy.

2.3. Hardness of austenite transformation products

A simple carbon equivalent C_{eq} is used here, and justified later:

$$C_{eq} = \mathrm{C} + \mathrm{Mn}/12 + \mathrm{Si}/24 \tag{11}$$

where element symbols refer to composition (wt%).

Cooling times that result in the formation of microstructures containing 50% martensite Δt_m^{50}, 0% ferrite Δt_f^0, and 0% bainite Δt_b^0 as a function of C_{eq} may then be obtained by regression analysis of published data for constructional steels [7]:

$$\Delta t_m^{50} = \exp(17.724\, C_{eq} - 2.926)$$

$$\Delta t_f^{\,0} = \exp(19.954\, C_{eq} - 3.944) \tag{12}$$

$$\Delta t_b^{\,0} = \exp(16.929\, C_{eq} + 1.453). \tag{13}$$

(14)

The volume fractions of the austenite transformation products martensite, bainite and a ferrite-pearlite mixture, V_m, V_b and V_{fp} respectively become (equation (7)):

$$V_m = \exp\{\ln(0.5).(\Delta t/\Delta t_m^{\,50})^2\} \tag{15}$$

$$V_b = \exp\{\ln(0.5).(\Delta t/\Delta t_b^{\,50})^2\} - V_m \tag{16}$$

$$V_{fp} = 1 - (V_m + V_b). \tag{17}$$

Regression analysis of published data for welds in constructional steels [7] gives the following equations for the average hardness of martensite, H_m, bainite, H_b and ferrite-pearlite, H_{fp}, in terms of the carbon equivalent used here:

$$H_m = 295 + 515\, C_{eq} \tag{18}$$

$$H_b = 223 + 147\, C_{eq} \tag{19}$$

$$H_{fp} = 140 + 139\, C_{eq}, \tag{20}$$

from which the average maximum HAZ hardness, $H_{max,}$ is calculated using the rule of mixtures:

$$H_{max} = V_m H_m + V_b H_b + V_{fp} H_{fp}. \tag{21}$$

3. Results and discussion

The chart in figure 1 is constructed using axes of the material index C_{eq} (on a linear scale) and the process index weld energy (on a logarithmic scale). It shows empirical composition ranges for steels and ranges of weld energy for various processes, and can be used as a scoping tool in a preliminary assessment of candidate materials and processes. However, no link is yet made between the indices.

The cooling curves commonly superimposed on CCT diagrams for steels are often labelled with empirically-determined volume fractions of transformation products. Figure 2 shows model predictions of the volume fractions of transformation products resulting from cooling curves that are characterised by different values of Δt. The variation of hardness with cooling time for a given steel, which may be calculated for a range of steel compositions using the methods described, is included schematically. The chart summarises the volume fractions of austenite transformation products based on the CCT diagram. It may be reconstructed for a different steel composition by using the models described. The variation of phase transformation products and maximum HAZ hardness may be assessed as a function of weld energy using this chart.

Figure 3 is constructed using the same axes as figure 1, scaled on the material axis for constructional steels, and shows model-based predictions of microstructural development in the HAZ after welding. The solid line bounds the region of martensite formation. Broken contours to the left of the boundary refer to conditions that result in a given volume fraction of martensite in the HAZ. Contours describing given volume fractions of bainite and a ferrite-pearlite mixture (not shown) may be constructed in a similar manner. By using a weldability criterion based on HAZ microstructure, e.g. 50% martensite, candidate combinations of welding parameters and steel composition may be assessed. By using equations (2) and (5) the weld energy scale may be modified to take into account preheating to a given temperature. The chart uses a material index and a process index to present solid state transformations in the HAZ of constructional steels, and is similar to the Schaeffler diagram, which shows weld metal microstructure in stainless steels following liquid-solid phase transformations using two material indices.

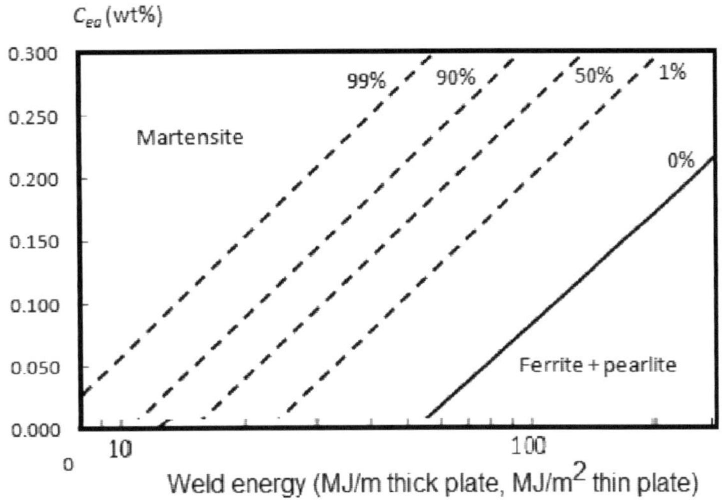

Figure 3. Chart showing model-based microstructural development in the heat affected zone of constructional steel welds. The axes are the process index (weld energy) and the material index (C_{eq}). The solid boundary shows the lower limit of martensite formation. Broken contours indicate a given volume fraction of martensite.

Figure 4 shows contours of constant maximum HAZ hardness, created using the model described for compositions relevant to constructional steels, and plotted using the axes of figure 1. Literature measurements of HAZ hardness adjacent to the fusion line (where the highest value is normally found) obtained from welds made using the manual metal arc, metal inert gas, submerged arc and laser processes [4, 8-10] are included. Experimental data are seen to agree well with the model-based hardness contours. By using a weldability criterion based on a maximum permitted hardness, such as that of 350 HV used in many branches of construction, an initial estimate of the welding parameters for safe welding of a steel of given composition may be obtained from figure 3. Alternatively the weldability of a range of steels within a given range of weld energy may be assessed rapidly.

Despite its simplicity, the material index C_{eq} appears to be a reliable measure of hardenability in constructional steels (model-based hardness contours lie in correct relation to experimental data on the y-axis of the chart). Similarly, the process index - weld energy input - appears to characterise the thermal cycles experienced in the HAZ well (experimental data and model predictions lie correctly on the chart x-axis).

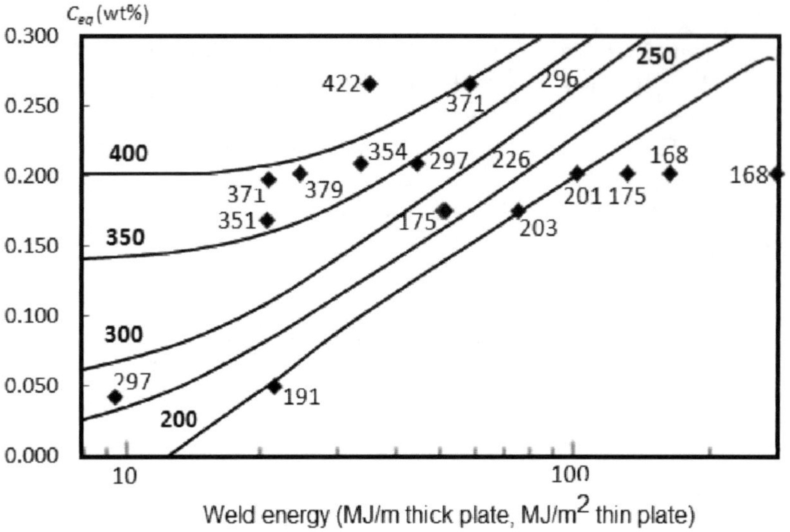

Figure 4. Chart showing hardness development in the heat affected zone of constructional steel welds. The axes are the process index (weld energy) and the material index (C_{eq}). Literature data for maximum heat affected zone hardness in welds produced by laser beam (thin plate), metal inert gas, tungsten inert gas and submerged arc (thick plate) processes are shown together with contours of hardness calculated using the model.

4. Conclusions

- The mathematical model used is a good predictor of the maximum hardness developed in the HAZ of constructional steels for a range of welding processes and steels.
- The charts are a user-friendly graphical display of the effects of variations in characteristic material and process indices on hardness and hardenability in constructional steels.
- The methods used provide a starting point to understand and model welding of steels currently under development, such as ausferritic and carbide-free grades.

5. References

[1] Ashby M F 2011 *Materials Selection in Mechanical Design* (Oxford: Butterworth-Heinemann)
[2] Ion J C, Ashby M F and Easterling K E 1984 *Acta Met.* **32** 1949
[3] Ion J C, Shercliff H R and Ashby M F 1992 *Acta Met. Mater.* **40** 1539
[4] Ion J C, Salminen A S and Sun Z 1996 *Weld. J.* **75** 225s
[5] Rosenthal D 1946 *Trans. ASME* **68** 849
[6] Johnson W A and Mehl R F 1939 *Trans. Am. Inst. Min. Met. Eng. Iron Steel Div.* **135** 416
[7] Inagaki M and Sekiguchi H 1960 *Trans. Nat. Inst. Met. Japan* **2** 102
[8] Albright C E, Hsu C and Lund R O 1991 Fatigue strength of laser-welded lap joints *Proc. Laser Materials Processing Conf. ICALEO '90* eds S L Ream, F Dausinger and T Fujioka (Orlando: Laser Institute of America) pp 357-363
[9] Moore P L, Howse D S and Wallach E R 2003 Microstructure and properties of autogenous high-power Nd:YAG laser welds in C-Mn steels *Proc. 6th Int. Conf. Trends in Welding Research* eds S A David, T DebRoy, J C Lippold, H B Smartt and J M Vitek, (Pine Mountain, GA, USA, 15-19 April 2002) (Materials Park: ASM International) pp 748-753
[10] Ion J C 2005 *Laser Processing of Engineering Materials* (Oxford: Elsevier)

Mechanically Alloyed Composite Materials Based on the Al-Mg-Li System Strengthened by Oxides

M Samoshina[1], **P Bryantsev**[1]
[1]National University of Science and Technology "MISiS", Leninskiy prospect, 4, Moscow, 119049, Russia

E-mail: samoshina@list.ru

Abstract. Mechanical alloying in air atmosphere lead to interaction of an initial material with the oxygen with formation of strengthening oxide particles. It allows creating the dispersion strengthened composite materials. The possibility of obtaining the composite materials based on Al-Mg-Li alloys by mechanical alloying is shown. It was found that the most intense oxidation of the material held within the first hour of milling. Microstructure evolution and mechanical properties of composite materials is investigated.

1. Introduction

It is known that mechanical alloying in chemically active atmosphere (for example, oxygen- or ammonia-containing) can lead to formation of new phases as a result of interaction of an initial material with the gas environment [1-3]. It allows creating the composite materials strengthened by disperse particles synthesized directly during mechanical alloying.

Magnesium and lithium in aluminum alloys have greater affinity for oxygen than aluminum and promotes the formation of complex oxides on the metal surface. Thus, in an oxygen atmosphere there is oxidation of Al, Mg and Li during mechanical milling of aluminum alloys containing magnesium and lithium. Therefore, Al-Mg-Li alloys are very promising for obtaining on their basis the composite materials strengthened by particles of natural oxides synthesized directly during mechanical milling.

2. Experimental

Dispersion-strengthened materials were obtained by mechanical milling of commercial chips of aluminum alloy Al – 5.5 wt.% Mg – 2 wt.% Li. The possibility of using the low-grade oxidized scrap such as turning and cutter chips to obtain quality composite materials has been shown in our earlier work [4]. The used chips were free of moisture and grease. Before treatment the chips were previously grinded to particles less 5000 μm in size.

Mechanical milling was performed in a vibratory mill SmV-0.005 in a constantly replenished air atmosphere for 1-15 h. It was assumed that strengthened oxide particles will be formed as a result of spontaneous and intensive oxidation of matrix alloy in contact with air. The air flow rate through the working space of the mill was ~ 0.4 m^3/h. The milling was produced by steel balls with diameter 12.5 mm. The ratio "milling balls:charge" was 30:1 in weight. The frequency of vibration was 30 Hz.

The granules after mechanical milling were consolidated to cylindrical samples of 15 mm in diameter and 10 mm in height by double-action pressing at a temperature of 400 ^{0}C.

Metallographic studies were performed on the light microscope Axiovision 200 MMAT. Measurement of granules size was carried out by method of random intercepts in several fields. The

IOP Conf. Series: Materials Science and Engineering **31** (2012) 012012 doi:10.1088/1757-899X/31/1/012012

total number of measured granules was 300-400. Microprobe analysis was performed with a scanning electron microscope «Camscan - 44» Link ISIS-200L by EDS method. This method may be used only for semiquantitative analysis of oxygen content. Structural analysis was carrying out by X-ray diffraction method on diffractometer D8 ADVANCE. The size of coherent scattering regions of aluminium solid solution was estimate by Williamson-Hall method.

The hardness and microhardness of the samples were estimated by the Vickers method. The hardness of consolidated samples was determined as the average of 10 measurements at the load of 5 kg and the holding time of 15 s. Microhardness measurements were carried out on the polished granules at the load of 10 g and holding time 10 s, measurements were made on 50 granules.

3. Results and Discussion

The morphology of the granules of composite materials after different milling times is shown in Fig. 1.

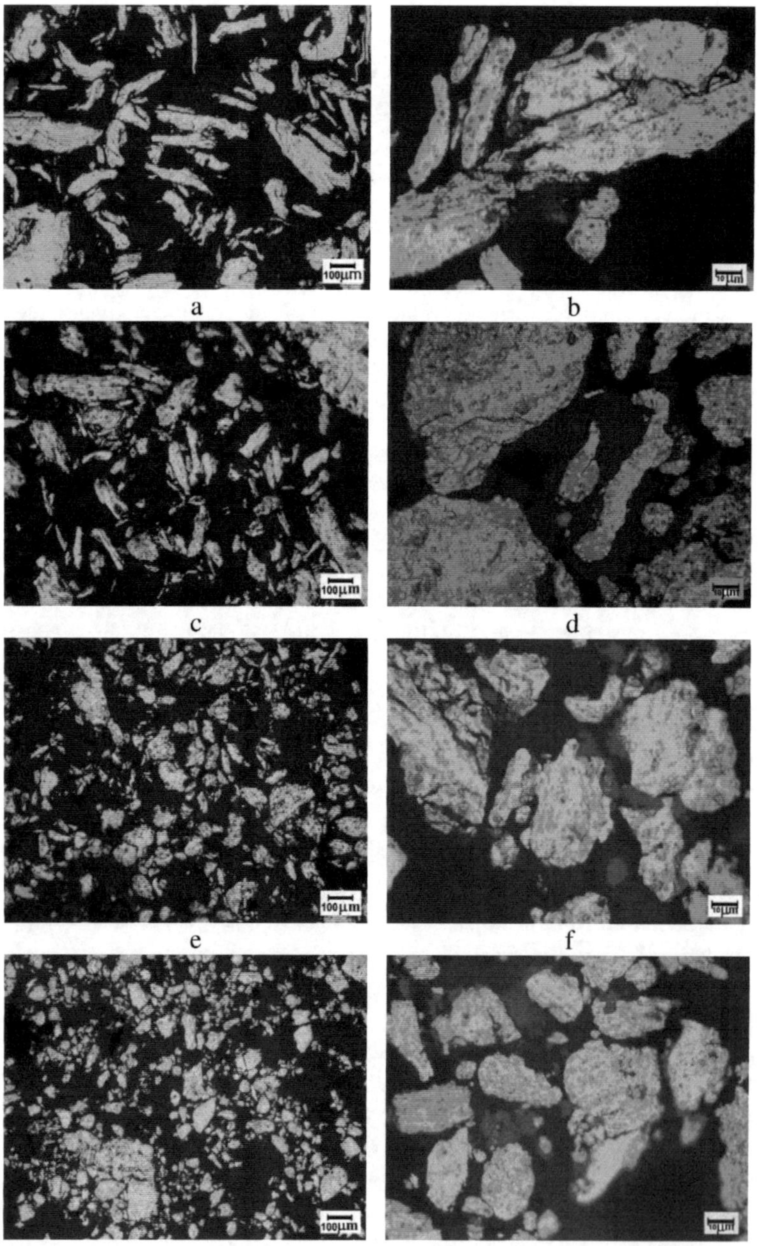

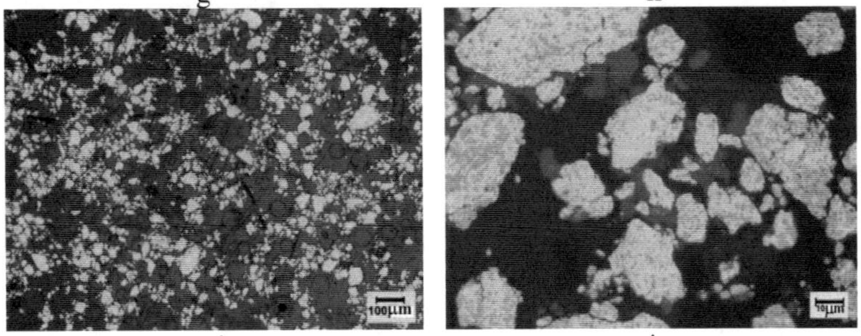

Fig. 1. Morphology of granules of composite material (Al – 5.5 % Mg – 2 % Li) – O after different milling time: a), b) 1 h; c), d) 3 h; e), f) 5 h; g), h) 10 h; i), j) 15 h.

For the initial milling time (1-3 h) the alloy particles have the form of flat flakes. After 5 h of milling granules begin to acquire equiaxial form. After 15 h of milling granules are heavily milled (fig. 2) and their form is nearly spherical.

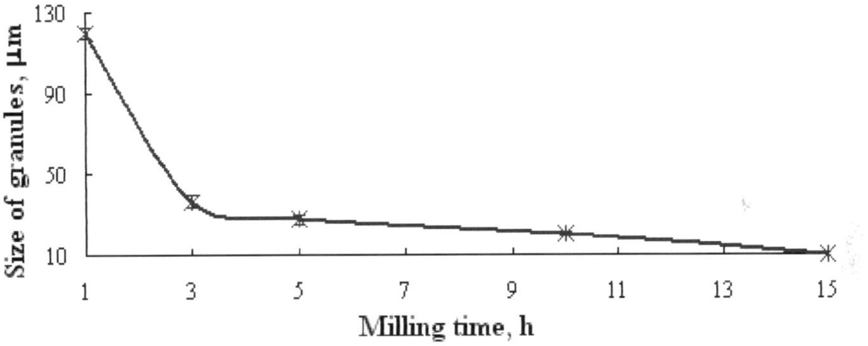

Fig. 2. Dependence of average size of granules of composite material (Al – 5.5 % Mg – 2 % Li) – O estimated by method of random intercepts on the milling time

We estimated the size of the coherent scattering regions (CSR) of aluminium solid solution , which corresponds to the size of grains or subgrains in the material (Al – 5.5 % Mg – 2 % Li) – O (fig. 3). Change of the size of the coherent scattering regions in granules of this material is due to alternation of work hardening and recovery and (or) recrystallization processes due to heating of the material during mechanical milling.

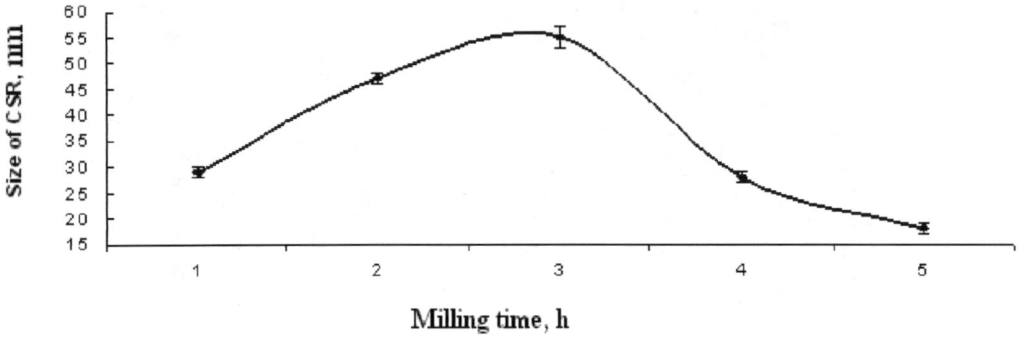

Fig. 3. Dependence of size of the coherent scattering regions of aluminium solid solution on the milling time

Fig. 4 shows the increasing of oxygen amount in the consolidated specimens determined by electron microprobe analysis depending on the time of milling in vibratory mill. It was found that the most intense oxidation of the material held within the first hour of milling, probably due to the rapid oxidation of magnesium and lithium in the matrix alloy.

X-ray phase analysis did not reveal the presence of any oxide particles in the specimen. This may indicates that the oxides particles have amorphous structure.

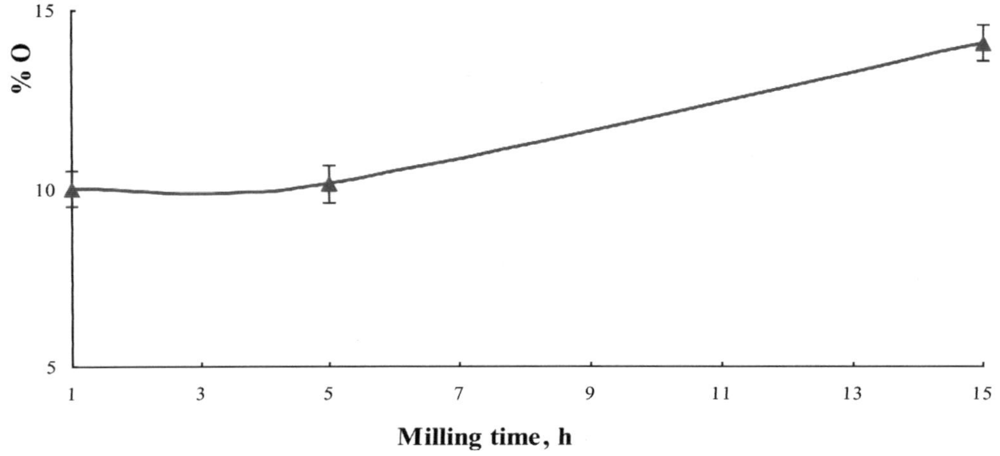

Fig. 4. Increase of oxygen amount in consolidated samples of composite material based on Al – 5.5 % Mg – 2 % Li alloy on the milling time.

Hardness is one of the simplest methods for determining the mechanical properties, which in this case also allows comparing the properties of granules and consolidated samples. Fig. 5 shows the hardness of granules and consolidated samples of the composite materials on the milling time.

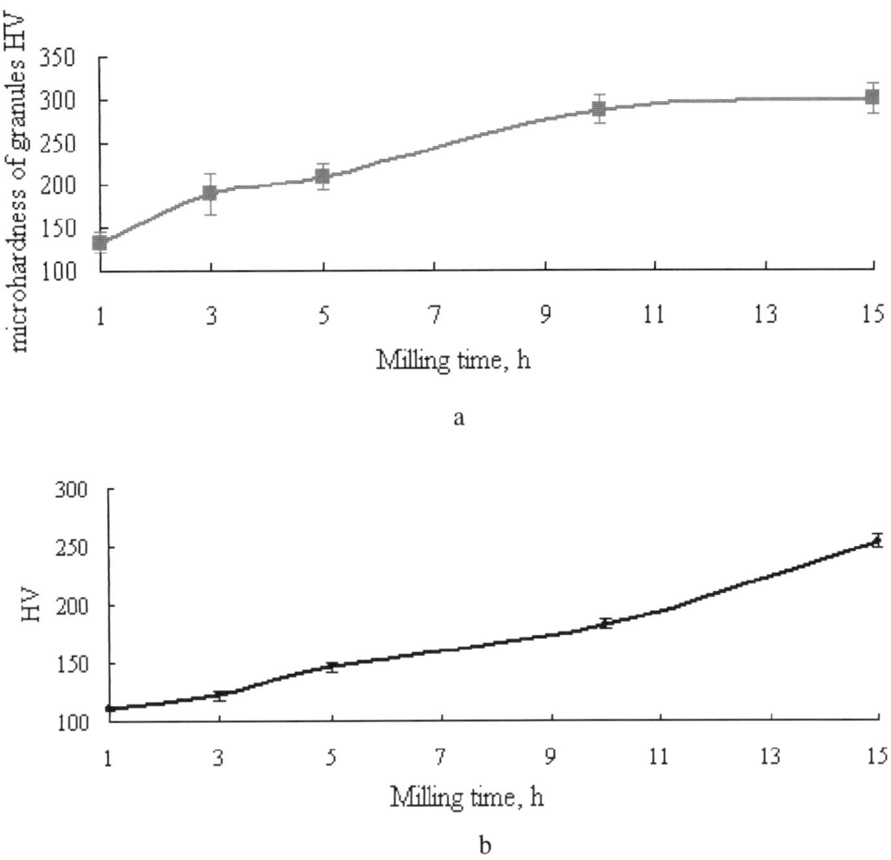

Fig. 5. Dependence of microhardness of granules of composite material on the milling time (a); dependence of hardness of consolidated samples of composite material on the milling time (b).

Hardness of the material increases significantly with the milling time due to the growth of the strengthening oxide particles amount. Reducing the hardness of consolidated samples compared with microhardness of granules is due to a general reduction of the defect concentration in structure during consolidation.

4. Conclusions

The possibility of obtaining the composite materials based on Al-Mg-Li alloys by mechanical alloying is shown. The formation of strengthening amorphous oxide particles during mechanical alloying in air atmosphere is confirmed. It was found that the most intense oxidation of the material held within the first hour of milling. Hardness of the composite material increases with the milling time due to the growth of the strengthening oxide particles amount.

Acknowledgments

The research is carried out with financial support of the Programme of Creation and Development of the National University of Science and Technology "MISiS" and Federal Program "Science personnel of innovative Russia" (state contract P1207 on 04.06.2010)

References

[1] P S Gilman and W D Nix 1983 *Metall. Trans. A.* **12** 813-884.
[2] R F Singer, W C Oliver and W D Nix 1980 *Metall. Trans. A.* **11** 1895-1901.

[3] J Cintas, F G Cuevas, J M Montes and E J Herrera 2005 *Scr. Mater.* **53** 1165–1170
[4] M Samoshina, A Aksenov and E Kaevitser 2008 *Rev. Adv. Mater. Sci.* **18** 305-310

Investigation of transformation of ferrous phases in 6XXX alloys during homogenizing annealing

P Bryantsev[1], M Samoshina[1]
[1]National University of Science and Technology "MISiS", Leninskiy prospect, 4, Moscow, 119049, Russia

E-mail: p_bryant@rambler.ru

Abstract. The only operation of heat treatment of 6XXX alloys before extrusion is homogenizing annealing. The presence of needle particles of β-Al_5FeSi in a billet for extrusion negatively affects the limit extrusion rate and surface quality of extruded products. Passage of the phase transformation of ferrous phases is the main limiting factor in the length of homogenizing annealing. In this work the method of quantitative phase transformation analysis is proposed for estimation the completeness of the homogenizing annealing.

1. Introduction

In recent years alloys of 6XXX series based on the system Al-Mg-Si took first place in production among all aluminum alloys. They make up more than 90% of the volume of the production of extruded aluminum products [1]. The reason why these materials are so popular is the excellent combination of such properties as high strength and corrosion stability and the possibility of depositing decorative and protective coatings by anodization and glazing. The extrusion rate of 6XXX series alloys may reach 50-70 m/min, an order of magnitude higher than the extrusion rate of other aluminum alloys.

The only operation of heat treatment of these alloys before extrusion is homogenizing annealing. In the process of homogenizing annealing, in addition to dissolution of the non-equilibrium particles of Mg_2Si phase, the transformation of needlelike β-Al_5FeSi phase to α-Al_8Fe_2Si occurs with more compact morphology. It is recognized that the presence of needle particles of β-Al_5FeSi in a billet for extrusion negatively affects the limit extrusion rate and surface quality of extruded products [1-5]. Passage of the phase transformation of ferrous phases is the main limiting factor in the length of homogenizing annealing.

2. Method of quantitative assessment of ferrous phases transformation

For accurate determination of the optimum temperature and annealing time of homogenization, it is necessary to have a quantitative criterion of ferrous phases transformation and changes in their morphology.

It should be noted that different methods were previously used to quantify the transformation of ferrous phases: light microscopy [6], scanning electron microscopy with microprobe analysis phase [7, 8], X-ray diffractometry [9, 10] and Mössbauer analysis [11].

In [12], Kuijpers conducted a comparative analysis of different methods. The direct identification of ferrous phases by electron diffraction was recognized as the most accurate method, but this method is very time-consuming, and the conclusions were only based on studies of 10-15 particles.

The needle-like morphology of β-Al5FeSi has a negative effect on extrusion process. To estimate the transformation of this ferrous phase in the process of homogenization, it is not required to carry out an accurate identification of the phase composition of the ferrous particles, but rather to trace the change in their shape. It is therefore very important to choose the parameter that describes the shape of the particles.

There are many mathematical parameters to describe the shape of complex objects. One of the most frequently used is the coefficient of elongation or the Feret ratio (aspect ratio), some studies applied it to study the phase transformation in aluminum alloys [6, 12]. This parameter is calculated as follows

$$\Phi = \frac{F_{\max}}{F_{\min}}, \qquad (1)$$

where F_{max} and F_{min} - the maximum and minimum Feret diameter respectively, i.e. minimum and maximum length of possible projections of the shapes on an arbitrary axis (Fig. 1a).

However, this parameter is not very suitable for determining the shape of curved objects. Fig. 1b shows a sample shape for which the value of F is close to 1, but the figure itself is far from equiaxed.

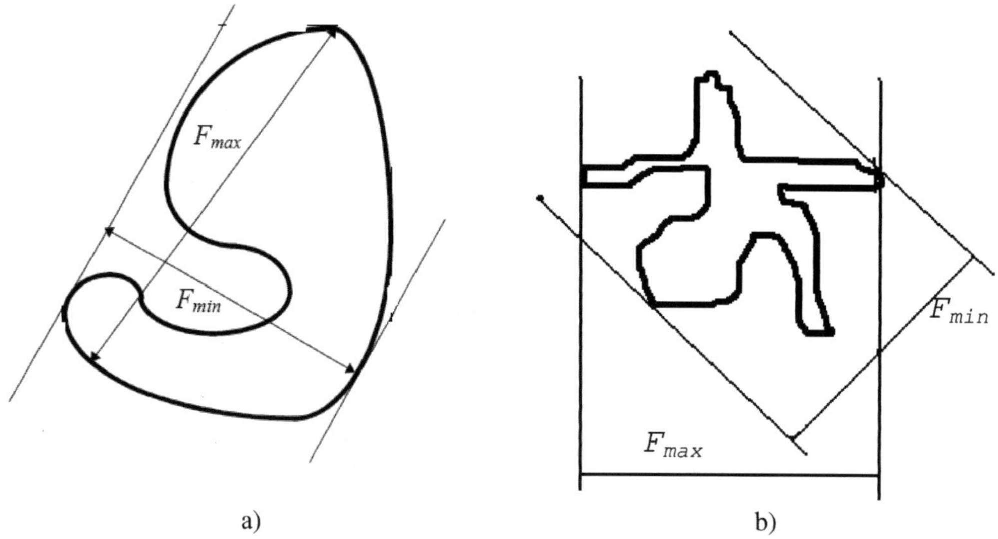

Fig. 1. The scheme for determining the coefficient of elongation F.

Quantitative assessment of transformation of ferrous phases was carried out using quantitative metallography software, Carl Zeiss Axiovision Rel. 4.5.

As the shape parameter of the particle automatically defined, function FormCircle (FC) was used. The latter is calculated as follows:

$$FC = 4\pi \frac{F}{P_C^2}, \qquad (2)$$

where F - the area of the particle;

P_C - Crofton perimeter of the particle, defined by the expression:

$$P_C = \frac{1}{4}\iint n(\varphi, p)d\varphi dp\,, \quad (3)$$

where $n(\varphi, p)$ – number of points on which the curve contouring particle intersects the line drawn at an angle φ and at a distance p from the origin.

From Crofton's theorem: the circle has the smallest area for a given perimeter. Therefore, the particles with spherical shape have the value of FC close to 1. Thus, we can determine the degree of fragmentation and spheroidization of the particles of ferrous phases, that is, evaluate the completeness of transformation.

3. Results and Discussion

Fig. 2 shows the microstructure of the alloy Al - 0.5 wt.% Mg – 0.5 wt.% Si – 0.2 wt.% Fe in as-cast state as well as a histogram of the distribution of ferrous phases particles in shape.

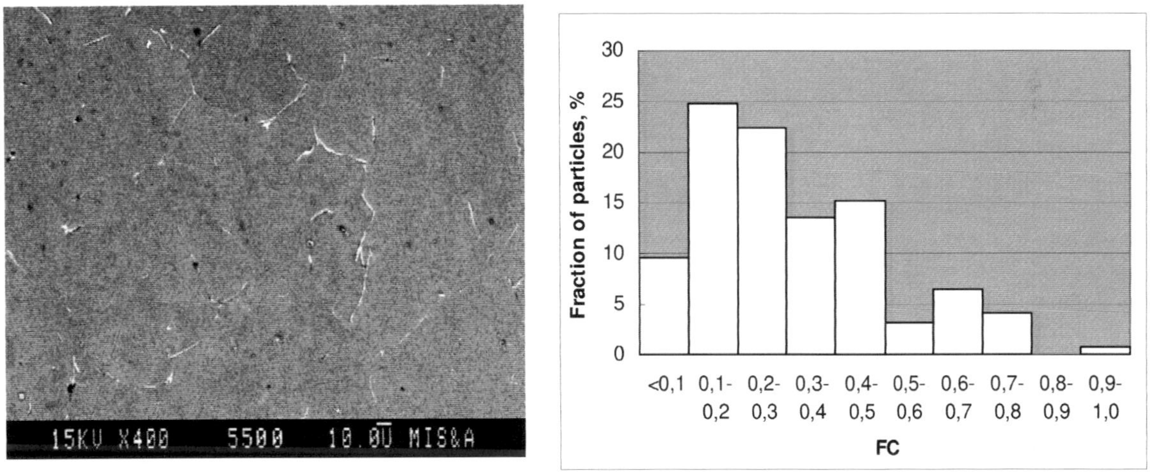

a b

Fig. 2. The microstructure of the alloy Al - 0.5 wt.% Mg – 0.5 wt.% Si – 0.2 wt.% Fe in as-cast state (a) and histogram of the distribution of ferrous phases particles in shape (b).

Fig. 2 shows that in as-cast state much of the ferrous particles has an elongated shape, so the analysis of the distribution of particles in the form shows that over 80% of the particles have a value of FC less than 0.5.

The results of quantifying the shape of the particles, after annealing at 580 ^{0}C, are shown in Fig. 3.

After holding at 580 ^{0}C for 4-6 hours, there is a significant fragmentation and spheroidization of ferrous particles - the fraction of particles with FC <0,5 is reduced almost 2 times compared with the as-cast state. With a 10-hour exposure, there is a further shift of the peak distribution towards higher FC values - almost 60% of the particles have FC in the range 0.6-0.9.

Figure 4 shows the histograms of phase distribution after annealing at 600 ^{0}C.

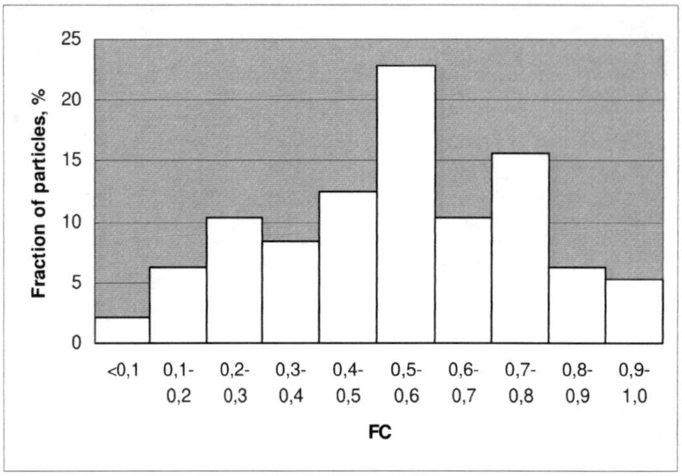

a

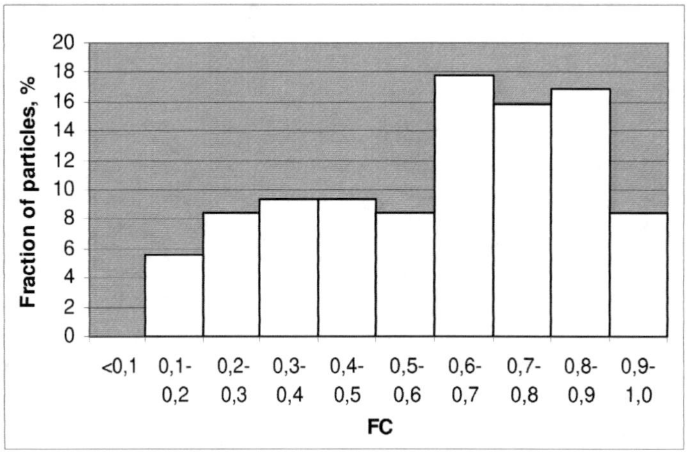

b

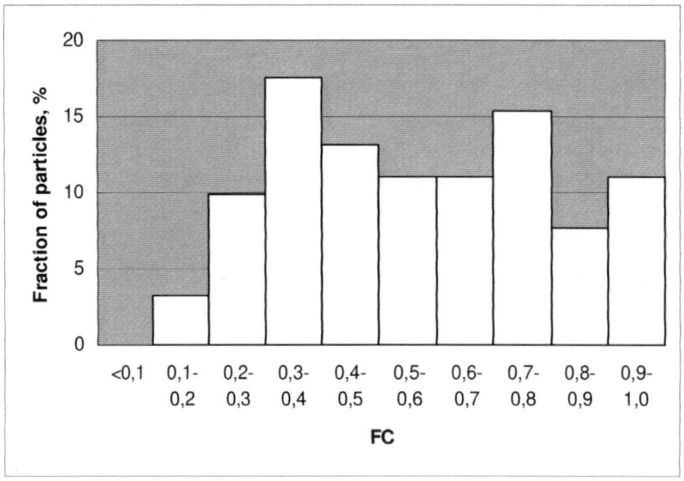

c

Fig. 3. Histograms of the distribution of ferrous phase particles in shape after annealing at 580 ^{0}C: 4 hours (a), 6 hours (b), 10 hours (c)

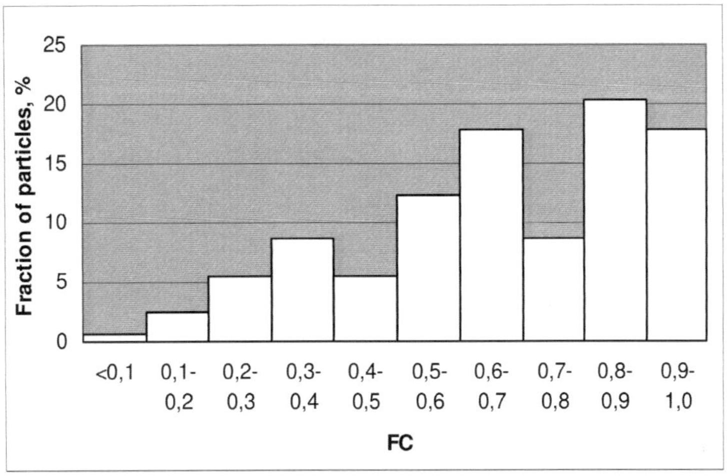

a

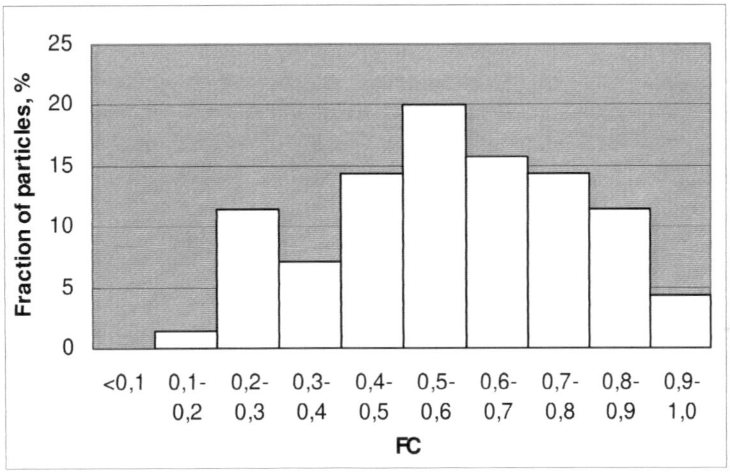

b

Fig. 4. Histograms of the distribution of ferrous phase particles in shape after annealing at 600 ^{0}C: 2 hours (a), 4 hours (b)

Already after annealing for 2 hours at 600 ^{0}C most elongated ferrous particles are disappeared. Only 20% of the particles have FC <0.5 after annealing for 4 hours.

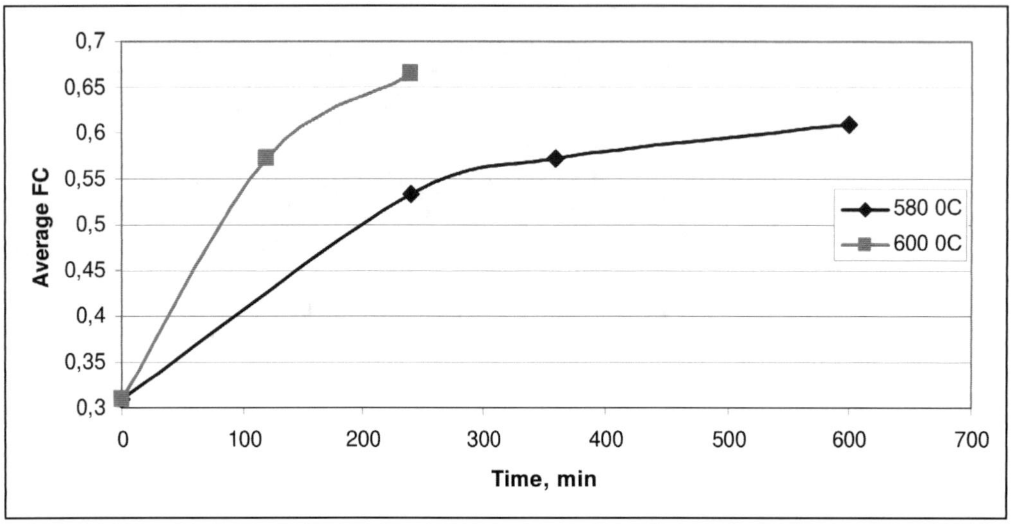

Fig. 5. Dependence of average FC on the annealing time and temperature.

Fig. 5. shows that the FC values continuously increase with increasing annealing time. This indicates the passage of spheroidization of ferrous particles. At the same time, increasing the annealing temperature accelerates the process of spheroidization. Fig. 5 shows that the annealing at 600 ^{0}C for 2 h leads to practically the same degree of spheroidization of ferrous particles as after annealing at 580 ^{0}C for 10 h. This is consistent with the observed microstructure.

Thus we can conclude that such a method of quantitative analysis is possible to be used for estimate the completeness of the homogenizing annealing.

Acknowledgments
The research is carried out with financial support of Grant of President of Russian Federation for young scientists MK-5561.2011.8 and the Programme of Creation and Development of the National University of Science and Technology "MISiS".

References

[1] O Reiso. Proc. of 9th Int. Conf. on Aluminium Alloys, ed. by J F Nie, A J Morton and B C Muddle 2004 Institute of Materials Engineering Australasia Ltd 32-46.
[2] N C W Kuijpers, F J Vermolen, C Vuik and S van der Zwaag. Proc. of 9th Int. Conf. on Aluminium Alloys, ed. by J F Nie, A J Morton and B C Muddle 2004 Institute of Materials Engineering Australasia Ltd 1040-1045.
[3] H Tanihata, T Sugawara, K Matsuda and S Ikeno 1999 *J Mater. Sci.* **34** 1205 – 1210
[4] S R Claves, D L Elias, W Z Misiolek. Proc. of the 8th Int. Conf. on Aluminium Alloys 2002 Trans Tech Publications Ltd 667-674.
[5] K B S Couto, M Goncalves, W H Van Geertruyden, W Z Misiolek. Proc. of the 9th Int. Conf. on Aluminium Alloys, ed. by J.F. Nie, A.J. Morton and B.C. Muddle 2004 Institute of Materials Engineering Australasia Ltd 1388-1393.
[6] Y Birol. 2004 *J Mater. Process. Technology* **148** 250–258
[7] K E Nilsen, P T G Koenis. Proc. of 7th Int. Aluminum Extrusion Technology Seminar 2000 Aluminum Extruders Council **2** 69–75.
[8] D T L Alexander, A L Greer. 2002 *Acta Mater.* **50** 2571– 2583.
[9] G Sha, K O'Reilly, B Cantor, J Worth, R Hamerton 2000 *Materials Science Forum* **304– 306** 253–258.

[10] T Minoda, H Hayakawa, H Yoshida. Proc. of 6th Int. Conf. on Aluminium Alloys 1998 The Japan Institute of Light Metals 339–344.
[11] S D Forder, J S Brooks, A Reeder, P V Evans. 1998 *Scr. Mater.* **40** 45–48.
[12] N C W Kuijpers, W H Kool, P T G Koenis, K E Nilsen, I Todd, S van der Zwaag 2003 *Materials Charact.* **49** 409– 420.

Investigation of the influence of copper welding electrodes on Ti-8Al-1Mo-1V and Ti-6Al-2Sn-4Zr-2Mo with respect to solid metal induced embrittlement

P Åkerfeldt[1], R Pederson[1,2] and M-L Antti[1]
[1]Division of Materials Science, Luleå University of Technology, S-97181 Luleå, Sweden
[2]Department of Materials Technology, Volvo Aero Corporation, S-46181 Trollhättan, Sweden

E-mail: pia.akerfeldt@ltu.se

Abstract. Solid Metal Induced Embrittlement (SMIE) is caused by a specific combination of two solid metals in intimate contact. Cadmium, gold, silver and copper are known to cause SMIE in certain titanium alloys. Solid copper is used in welding electrodes and fixtures in various manufacturing processes for titanium parts within the aerospace industry. In the case of resistance welding, titanium alloys are in intimate contact with solid copper, since the electrodes resistively heat the titanium part under pressure during the welding process. No previous published work that investigates the risk of using copper electrodes for welding of titanium alloys is available in the literature, but an initial study using U-bend testing indicates that solid copper in contact with Ti-8Al-1V-1Mo and Ti-6Al-2Sn-4Zr-2Mo could lead to SMIE. Therefore, in the present study, resistance welded Ti-8Al-1V-1Mo and Ti-6Al-2Sn-4Zr-2Mo have been evaluated to investigate the influence of copper electrodes on these alloys. Furthermore, resistance welded specimens sputtered with copper and gold to promote SMIE have also been evaluated. No SMIE was found in the resistance welded specimens, which may be explained by the short interaction time that the copper electrodes are in intimate contact with the titanium alloy, and/or the magnitude of residual stresses after welding, which may be too low to initiate SMIE.

1. Introduction

One of the joining methods that can be used for titanium components within the aerospace industry is seam welding. In seam welding the welding electrodes, commonly made of copper, press overlapping work pieces together and heat them to the welding temperature to form one of the nuggets in a seam weld [1]. However, there are doubts about the influence of the copper electrodes on the titanium work piece; recently the authors found that two near alpha alloys were susceptible to solid metal induced embrittlement (SMIE) in contact with solid copper [2]. For SMIE to occur, two solid metals must be in intimate contact with each other [3-8], which is the case for the copper electrodes and the titanium part during the seam welding process, and which could be the case when the copper electrode degrades if wear debris becomes attached to the titanium part. However, there are other important parameters as well, e.g. the temperature, holding time and surface condition. The mechanism of SMIE is believed to be a combination of surface self-diffusion of the embrittling species to the crack tip and adsorption of the embrittling species at the crack tip [3-5], see figure 1. Surface self-diffusion is assumed to be the

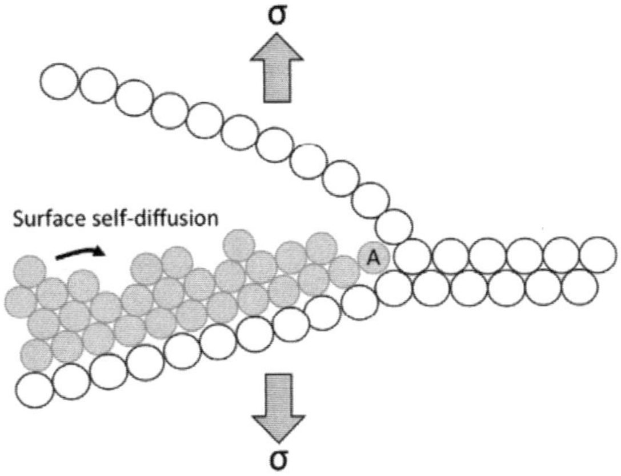

Figure 1. Illustration of surface self-diffusion of embrittling metal atoms to crack tip and adsorption of atom "A" at crack tip under the tensile stress σ.

transport mechanism since its diffusion rates are fast enough to be consistent with cracking lengths at the temperatures studied [5,6,9-10] compared with vacancy and grain boundary diffusion. Concerning how the adsorption affects the crack tip, the details are not fully understood but in general it is agreed that the adsorption weakens the crack tip. The two main theories are (1) adsorption reduces the stress required for tensile separation of the atom at the crack tip leading to decohesion [8,11-13] and (2) adsorption facilitates the nucleation or egress of dislocations at crack tip [5, 14-16].

The mechanism of SMIE susceptibility is rather complex to model, but experimental studies indicate that cracking resulting from SMIE is expected to occur above a homologous temperature (T/T_m) of 0.5 [6], where T is the service temperature and T_m the melting temperature of the embrittling species (K). For titanium alloys, SMIE has been observed in contact with the embrittling species listed in table 1.

Table 1. Titanium alloys and embrittling species reported.

Alloy	Embrittling species
Ti-3Al-14V-11Cr	Cd [3,17]
Ti-6Al-4V	Cd [17,18]
Ti-6Al-6V-2Sn	Ag [19], Au [19], Cd [19]
Ti-6Al-2Sn-4Zr-2Mo	Au [2], Cu [2]
Ti-6.5Al-3.5Mo-1.5Zr-0.3Si	Cu [20]
Ti-8Al-1Mo-1V	Au [2], Cd [17,18], Cu [2]

In the present paper, the influence of copper welding electrodes on two titanium alloys has been evaluated and discussed. Although failures by SMIE are rare in industry, it does occur, partly because of the lack of awareness of SMIE. This study is made in close collaboration with the aerospace industry to increase understanding and knowledge of SMIE.

2. Materials and method

2.1. Materials
Two near alpha alloys were evaluated; Ti-8Al-1Mo-1V and Ti-6Al-2Sn-4Zr-2Mo, both of which have been shown to be susceptible to SMIE in contact with copper and gold in a previous study [2]. The microstructure of the near alpha alloys investigated mainly consists of alpha grains surrounded by a small amount of beta phase. This kind of microstructure is typical in sheet material of these alloys. The condition and composition of the sheets evaluated here are in accordance with AMS4916J [21] (Ti-8Al-1Mo-1V) and AMS4919F [22] (Ti-6Al-2Sn-4Zr-2Mo).

2.2. Welding procedure
The sheets were seam welded by using resistance welding equipment with wheel-shaped electrodes, made of copper, see figure 2(a). The seam weld can be divided into nuggets that represent one welding cycle, as shown in the cross-section of a seam weld (the nugget) in figure 2(b). A welding cycle comprises the time for which the electrodes are in contact with the work piece, including the set-up time (the time taken to achieve the correct mechanical pressure) and the cooling time. In the present study, one nugget was made using two weld impulses of 160 ms, with a total welding cycle of 2160 ms. The sheets were cut to 75x19x1.25 (±0.25) mm parallel to the rolling direction, and the seam weld was made in the middle of the sheet, see figure 2(c).

2.3. Test procedure
The main parts of the test procedure were adapted from SAE ARP 1795A [23], which is a standard U-bend test method normally used to evaluate stress-corrosion cracking of titanium alloys caused by aircraft cleaning materials. However, the seam welded specimens were not U-bent since the aim was to evaluate the residual stresses from the weld itself. To avoid contamination or embrittlement caused by other sources, the specimens were immersed in a cleaning solution containing 35vol% nitric acid,

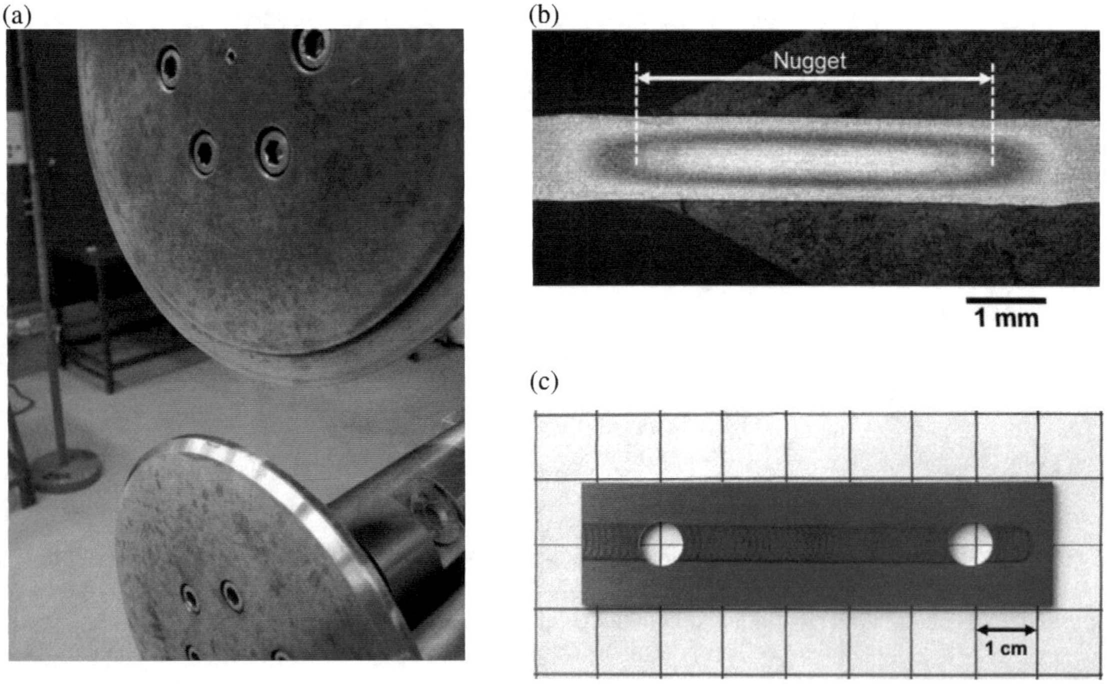

Figure 2. (a) Wheel-shaped copper electrodes, (b) cross-section of a seam weld etched with Kroll's reagent showing the nugget, and (c) typical appearance of seam-welded sheets.

3vol% hydrofluoric acid and reagent water for 20 seconds, then rinsed in reagent water and left to air dry. The specimens were thereafter carefully handled with cotton gloves throughout the experimental procedure. Three conditions were evaluated: seam welded specimens sputtered with either i) copper or ii) gold, and iii) specimens that had only been seam welded. The intention with the sputtered copper and gold was to promote embrittlement by guaranteeing the presence of embrittling species. Both copper and gold have, in previous studies, been shown to embrittle titanium alloys, see table 1. Depending on the sputtering time, current and type of solid metal, the thickness of the sputtered layer was between 50-100 nm.

Table 2. Alloys tested, conditions and heat treatments.

Alloy	Condition	Temperature [°C] / Duration [hours]
Ti-8Al-1Mo-1V	Seam welded	480 / 8
		593 / 2
	Seam welded + copper	480 / 8
		593 / 2
	Seam welded + gold	480 / 8
		593 / 2
Ti-6Al-2Sn-4Zr-2Mo	Seam welded	480 / 8
		593 / 2
	Seam welded + copper	480 / 8
		593 / 2
	Seam welded + gold	480 / 8
		593 / 2

The subsequent heat treatment was carried out according to two procedures. One batch of specimens was heat treated at 480°C for 8 hours and acid pickled according to SAE ARP 1795A [23]. The other batch was heat treated at 593°C for 2 hours and acid pickled according to a process route used in the aerospace industry. A summary of the alloys evaluated, conditions and heat treatments can be found in table 2.

Following heat treatment and acid pickling, all specimens were carefully cut along and across the seam weld, as indicated by the arrows in figure 3, to produce two cross-sections along the weld length

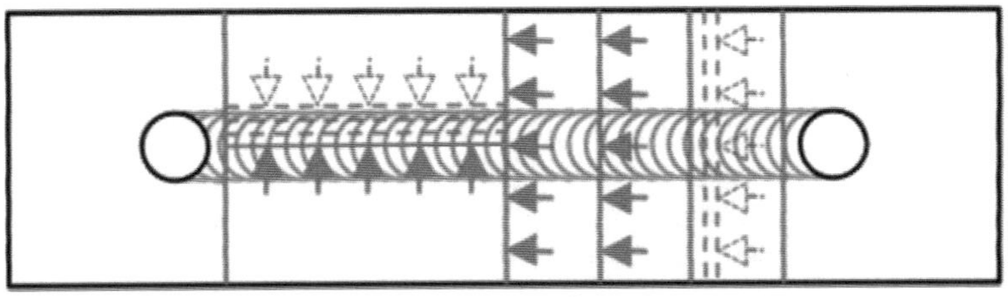

Figure 3. Schematic sketch of the cross-sections evaluated. The dashed lines and arrows illustrate the locations of the re-ground and re-polished cross sections.

and three cross-sections across the weld width. The cross-sections were mounted, ground and polished using conventional methods for titanium alloys. The evaluation of the cross-sections was carried out in a light optical microscope at 500x. If no crack was observed, the mounts were reground and re-polished to a depth of approximately 1 mm, as indicated by the dashed lines and arrows in figure 3.

3. Results and discussion

No indication of SMIE caused by the copper electrodes was observed in seam welded Ti-8Al-1Mo-1V or Ti-6Al-2Sn-4Zr-2Mo, neither was any SMIE observed in the specimens sputtered with copper and gold. All specimens were re-ground, re-polished and re-examined several times without displaying any indication of SMIE.

If one considers the homologous temperature (T/T_m) above which SMIE is expected to occur (0.5), SMIE caused by copper has a threshold temperature of 406°C. During seam welding the contact surface of the copper electrodes was heated to approximately 900°C (T/T_m = 0.86), i.e. to a much higher temperature than the threshold temperature. However, the electrodes were in contact with the titanium work piece for a very short time (320 ms), resulting in negligible diffusion. During the welding operation, the copper electrodes also degrade, resulting in wear debris, which could attach to the surface of the titanium work piece - this was experimentally studied by sputtering copper onto the surface of the test specimens. After welding, the sheets are normally stress relieved and so the test specimens in the present study were also stress relieved. In this study, the stress relieving treatments were performed at 480°C (T/T_m=0.55) for 8 hours or at 593°C (T/T_m=0.64) for 2 hours depending on the batch. The diffusion that occurs during heat treatment can be estimated by calculating a characteristic diffusion length. The surface self-diffusion coefficient (D_s) for copper depends strongly on the surrounding atmosphere; in an oxygen atmosphere the pre-exponential diffusion coefficient (D_0) is 0.15 cm^2/s, and the activation energy (Q) is 75 kJ/mol [24]. The surface self-diffusion coefficients (D_s) at 480°C and 593°C are thus calculated to be 9.4×10^{-7} and 4.5×10^{-6} cm^2/s, respectively using a form of the Arrhenius equation [25]:

$$D_s = D_0 \exp(-Q/RT) \qquad (1)$$

A characteristic diffusion length may then be calculated using the random-walk diffusion equation [4-5, 18]:

$$\bar{x} = \sqrt{2D_s t} \qquad (2)$$

where t is time (s). Assuming that the oxygen atmosphere is a valid approximation for air, the present holding times (t) would result in characteristic diffusion lengths of 0.0023 m (480°C, 8h) and 0.0025 m (593°C, 2h), which are sufficient to cause severe embrittlement.

Whilst no cracking was observed in the seam welded specimens, cracking was observed after 480°C for 8 hours in the U-bend test [2] for both Ti-8Al-1Mo-1V and Ti-6Al-2Sn-4Zr-2Mo that had copper or gold sputtered onto the surfaces. One of the prerequisites for SMIE is tensile stress, which in this study is expected to be generated by the seam welding process itself. Since the work piece is heated to the welding temperature, the material undergoes thermal expansion, which is restrained by cooler adjacent material, resulting in both tensile (in the weld) and compressive (adjacent to the weld) residual stresses. The stress distribution may be different in the seam weld since the weld consists of several small spot welds. In addition, the distribution and size of the residual stresses are also dependent on the alloy in question, phase transformations that occur, cooling rate, peak temperature, and grain size [26] and are therefore difficult to predict. In the present study, the seam welded specimens were evaluated in two orientations to cover at least two possible crack growth directions -

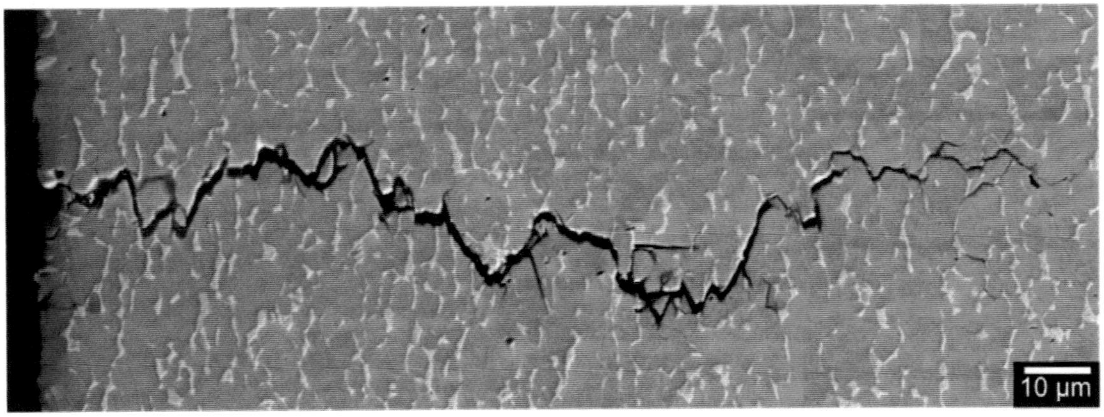

Figure 4. SMIE observed in Ti-6Al-2Sn-4Zr-2Mo in contact with copper, showing cracking located in the low-stress area at the legs of the U-bend shape [2].

along and across the seam weld - see figure 3. However, even if the residual stresses induced by welding attain the yield stress in some cases, it is supposed that the stress levels in the present study were less than those observed in previous U-bend tests [2] since the sheets in the U-bend tests experience plastic deformation as they are bent. It should on the other hand be noted that in the U-bend test, severe cracking was observed in the low stress region at the legs of the U-bend in Ti-6Al-2Sn-4Zr-2Mo in contact with copper, see figure 4, indicating that SMIE may also occur at stress levels lower than the yield stress.

Titanium alloys are known for their ability to form a stable protective oxide surface layer when exposed to air [27]. In this study, the oxide layer was largely removed by acid pickling prior to sputtering and/or heat treatment, but since titanium alloys are highly reactive a new oxide layer is instantly formed when exposed to air. Thus, during intimate contact between the two solid metals, for crack initiation to occur the oxide must be ruptured either mechanically or chemically. This also emphasises the role of tensile stress, which can tear apart the oxide film and maintain intimate contact of the embrittling species with the exposed titanium at the crack tip. The inhibition of transport and adsorption processes of the embrittling species because of the oxide layer has been observed by Fager and Spurr [18], who studied SMIE of titanium alloys in contact with cadmium. The oxide layer could explain why no embrittlement was observed for the seam welded specimen, i.e. the tensile stresses were too low. Further, SMIE in the U-bend specimens may be explained by a higher stress level, even when compared with the low stress region in the legs of U-bend samples. In addition to the threshold temperature, a threshold stress level above which SMIE is more likely to occur is believed to exist.

Concerning the applicability of seam welding of titanium alloys in the aerospace industry, there are additional parameters to consider before establishing the influence of copper electrodes or copper debris on SMIE in titanium alloys. First, one must consider the influence of additional forces that could increase the tensile stress and thereby increase the susceptibility to SMIE. Secondly, cyclic forces must be considered; tests have so far only been carried out with static forces.

4. Conclusions

No cracks were found in the seam welded specimens, which could be explained by the absence of copper in the weld. However, since no cracks were found neither in the seam welded specimens sputtered with copper or gold, it can be concluded that the absence of copper was not the reason for the absence of embrittlement in the present study. Therefore, it can be concluded that the residual stresses generated by the seam welding operation itself are insufficient to induce cracks through SMIE in Ti-8Al-1Mo-1V and Ti-6Al-2Sn-4Zr-2Mo.

References

[1] Dent P, Bohr J C, Gasser R G, Gerken J M, Hallum D L, Lee J W, McCauley R B, Orts D H, Oyler G W, Shieh W T and Wu K C 1991 Spot, Seam, and Projection Welding *Welding Processes (Welding Handbook* vol 2) ed R L O'Brien (Miami: American Welding Society) chapter 17 pp 531-79

[2] Åkerfeldt P, Pederson R and Antti M-L 2011 Solid metal induced embrittlement of titanium alloys in contact with copper *Proc. of The 12th World Conf. on Titanium (Ti-2011)* (Beijing) Accepted for publication

[3] Kamdar M H 1988 Solid metal induced embrittlement *J. Electrochem. Soc. India.* **37** 53-9

[4] Lynch S P 1992 Metal-induced embrittlement of materials *Mater. Charact.* **28** 279-89

[5] Lynch S P 1989 Solid-metal-induced embrittlement of aluminium alloys and other materials *Mater. Sci. Eng. A* **108** 203-12

[6] Lynch S P 2008 Failures of structures and components by metal-induced embrittlement *J. Fail. Anal. and Preven.* **8** 259-74

[7] Lynch S P 1994 Failures of structures and components by environmentally assisted cracking *Eng. Failure Anal.* **1** 77-90

[8] Kamdar M H 1983 Liquid Metal Embrittlement *Embrittlement of Engineering Alloys* (*Treatise on Material Science and Technology* vol 25*)* ed C L Briant and S K Banerji (London: Academic Press) chapter 9 pp 362-455

[9] Gordon P 1977 Metal induced embrittlement of metals – an evaluation of embrittler transport mechanisms *Metallurg. Trans. A* **9** 267-73

[10] Lynn J C Warke W R and Gordon P 1975 Solid metal-induced embrittlement of steel *Mater. Sci. Eng. A* **18** 51-62

[11] Westwood A R C and Kamdar M H 1963 Concerning liquid metal embrittlement, particulary of zinc monocrystals by mercury *The Philosophical Magazine* **8** 787-804

[12] Westwood A R C, Preece C M and Kamdar M H 1967 Application of crack propagation criterion to liquid-metal embrittlement; cleavage of aluminium monocrystals in liquid gallium *Trans. Am. Soc. Metals* **60** 723-5

[13] Stoloff N S and Johnston T L 1963 Crack propagation in a liquid environment *Acta Metallurg.* **11** 251-6

[14] Lynch S P 1984 Metal-induced embrittlement of ductile materials and dislocation emission from crack tips *Scripta Metallurgica* **18** 509-13

[15] Lynch S P 1978 Effect of environment on fracture – mechanisms of liquid-metal embrittlement, stress-corrosion cracking and corrosion-fatigue *Advances in research on the strength and fracture of materials* (Waterloo)

[16] Lynch S P 1979 Mechanisms of fatigue and environmentally assisted fatigue *Fatigue Mechanisms* ed Fong J T (Philadelphia: American society for testing materials) pp 175

[17] Meyn D A 1973 Solid cadmium cracking of titanium alloys *Corrosion* **29** 192-6

[18] Fager D N and Spurr W F 1970 Solid cadmium embrittlement: titanium alloys *Corrosion* **26** 409-19

[19] Stoltz R E and Stulen R H 1979 Solid metal embrittlement of Ti-6Al-6V-2Sn by cadmium, silver and gold *Corrosion* **35** 165-9

[20] Liu D -X, Liu S -M and Fan G –F 2006 Investigation on behavior of solid copper-induced embrittlement of titanium alloy *J. Aeronaut. Mater.* **26** 1-5

[21] AMS 4916J 2006 Aerospace Material Specification: Titanium alloy sheet, strip, and plate 8Al-1Mo-1V duplex annealed (Warrendale: SAE International)

[22] AMS 4919F 2006 Aerospace Material Specification: Titanium alloy sheet, strip, and plate 6Al-2Sn-4Zr-2Mo-0.08Si duplex annealed (Warrendale: SAE International)

[23] SAE ARP 1795A 1984 Stress corrosion of titanium alloys: effect of cleaning agents on aircraft engine materials (Warrendale: SAE International)

[24] Butrymowicz D B, Manning J R and Read M E 1973 Diffusion in copper and copper alloys *J. Phys. Chem. Ref. Data* **2** 643-55
[25] Porter D A and Easterling K E 1981 *Phase transformations in metals and alloys* (Wokingham: Van Nostrand Reinhold Co Ltd)
[26] Easterling K 1983 *Introduction to the physical metallurgy of welding* (Oxford: Butterworth Heinemann)
[27] Lütjering G and Williams J C 2003 *Titanium* (Berlin Heidelberg: Springer-Verlag)

Acknowledgements
The authors would like to acknowledge Claire Karlsson for her assistance in the project, Dr. John Ion for all helpful discussions, Volvo Aero Corporation for financial support and collaboration, the Graduate School of Space Technology for the encouragement, and the National Aviation Engineering Research Program (NFFP) for funding the project.

IOP Conf. Series: Materials Science and Engineering **31** (2012) 012015 doi:10.1088/1757-899X/31/1/012015

Surface mechanical properties of advanced zirconia after hydrothermal exposure

F G Marro[1,2], A Mestra[1,2], E Jiménez-Piqué[1,2], R Chintapalli[1], M Anglada[1,2]
[1]Dept. of Materials Science and Metallurgical Engineering, Universitat Politècnica de Catalunya, Av. Diagonal 647, 08028 Barcelona, Spain
[2]Centre for Research in Nanoengineering, CRnE, Universitat Politècnica de Catalunya, Pascual i Vila 15, 08028 Barcelona, Spain

E-mail: fernando.garcia.marro@upc.edu

Abstract. The surface mechanical properties of different types of advanced zirconia ceramics have been assessed after being subjected to hydrothermal degradation. Three different types of zirconia were considered: standard tetragonal polycrystalline zirconia doped with 3 % molar yttria (3Y-TZP) produced by conventional sintering; 3Y-TZP produced by spark-plasma sintering (3Y-TZP(SPS)) and a Ce-TZP/Al_2O_3 nanocomposite. Hydrothermal ageing was assessed by using X-ray diffraction. Berkovich nanoindentation was performed before and after the materials being exposed to 131 ºC water vapour in autoclave, in order to assess changes in the surface hardness and Young modulus. It is shown that, while standard 3Y-TZP suffers a substantial decay of the surface mechanical properties with hydrothermal exposure, this is not the case for 3Y-TZP(SPS) and the Ce-TZP/Al_2O_3 materials. On the contrary, the later maintained their surface integrity during hydrothermal exposure. The results are explained in terms of microstructure and chemical composition.

1. Introduction

Tetragonal polycrystalline zirconia doped with 3 mol % of yttria (3Y-TZP) is a ceramic with industrial relevance due to its good mechanical resistance of around 1000 MPa, a moderate fracture toughness of 5 MPa·m$^{1/2}$, a good wear resistance and excellent biocompatibility. These properties have made this ceramic interesting for medical applications like balls for femoral heads and recently in dentistry. However, 3Y-TZP is known [1]-[4] to be susceptible to degradation in the presence of a humid environment in a range of temperatures between 25-300 ºC. In this process, the exposure to water for a sufficiently long period of time promotes the formation of a surface layer of monoclinic phase which is accompanied of microcracks and has lower mechanical integrity than the bulk. This phenomenon, usually referred to as hydrothermal ageing or low temperature degradation, has been one of the main causes for in vivo failure of 3Y-TZP femoral heads several months after implantation. Although it was later demonstrated [4] that the cause for hydrothermal degradation of the failed heads was the use of incorrect processing parameters, hydrothermal ageing has remained nevertheless a serious issue for any biomedical application.

In order to avoid this important effect, in recent years there has been many efforts to develop zirconia based ceramics which do not suffer ageing in humid environments while maintaining the good mechanical properties of 3Y-TZP. One of the most important strategies for doing so has been the development of zirconia-alumina nanocomposites [5]-[6]. Another strategy to diminish the susceptibility to ageing consists in reducing the size of the final microstructure by new processing techniques like spark plasma sintering (SPS), as the grain size affects the susceptibility to ageing.

The very low susceptibility to ageing of the ceramics developed by these strategies has been demonstrated by X-ray diffraction [7]-[8]. By contrast, here we focus the attention on the surface mechanical properties after these materials have been exposed to accelerated hydrothermal ageing by using hot water vapour in autoclave. Instrumented nanoindentation is used in the present work to extract mechanical properties of the surface. The potential of this technique for detecting changes in surface mechanical properties has been proved recently in the study of degraded 3Y-TZP [9].

2. Materials and Methods

Three materials were considered: standard as sintered 3Y-TZP ("AS"); 3Y-TZP produced by spark plasma sintering (3Y-TZP(SPS)) and a Ce-TZP/Al_2O_3 nanocomposite. The AS material was produced from commercial TZ-3YSB-E powder (Tosoh Co., Japan) which was cold isostatically pressed at 200 MPa producing green compact bars which were sintered in an alumina tube furnace at 1450 °C for 2 hours with heating and cooling rates of 3 °C/min. The 3Y-TZP(SPS) was produced from the same powder but it was sintered in a SPS system (SPS FCT HP D25I, FCT system GmBh) at temperatures of 1200 °C for a soak time of 5 minutes at a pressure of 100 MPa. The density was measured by the Archimedes method and was over 99 % of the full density theoretical value, the bulk hardness was measured by 1 kg Vickers indentation and the indentation fracture toughness was determined from the dimensions of the cracks produced by a 30 kg indentation. These cracks were of Palmqvist type (not connected below the surface) and from their length the fracture toughness was obtained with the following equation [10]:

$$K_{IC} = 0.025(E/H)^{0.4}\sqrt{\frac{PH}{4l}} \qquad (1)$$

Where P is the indentation load, l the crack length at the surface and the hardness parameter stands for $H=P/2a^2$ where a is the imprint semi-diagonal. In general, indentation methods give only rough estimations of the actual fracture toughness; however these methods are known to actually give values close the true fracture toughness for standard 3Y-TZP. On the other hand, the fracture toughness of the Ce-TZP/Al_2O_3 nanocomposite cannot be analyzed by this method because of the large transformation around the indentation imprint so that other methods have to be used. The fracture toughness has been determined in a previous work for the present nanocomposite [8]. A resume of all these properties for the ceramics used here is given in table 1. Additionally, the microstructure was observed by scanning electron microscopy after polishing the samples up to colloidal silica and thermal etching.

Table 1. Density, average grain size, Vickers hardness and fracture toughness of the three ceramics considered.

Sample	Density ($g\cdot cm^{-3}$)	HV1 (GPa)	K_{IC} ($MPa\cdot m^{1/2}$)
AS	6,06	12.0 ± 0.1	5.1 ± 0.1
3Y-TZP(SPS)	6,00	14.1 ± 0.7	3.4 ± 0.4
Ce-TZP/Al_2O_3	5,53	11.0 ± 0.6	10.0 ± 0.5

The specimens were subjected to hydrothermal ageing by exposure to 131 °C water vapour in autoclave at a constant pressure of 2 bars for a period of 100 h. From previous studies, this test is known to be a sufficiently aggressive test to completely transform the surface of the standard 3Y-TZP into monoclinic phase. X-ray diffraction was performed with Cu K_α incident radiation (1.5406 Å) in symmetric θ-2θ configuration with a Bruker AXS D8 diffractometer. The monoclinic volume fraction on the specimens subjected to ageing was determined from the diffraction spectra with the following equation [11]:

$$V_m = \frac{1.311\left(I_m(\bar{1}11) + I_m(111)\right)}{1.311\left(I_m(\bar{1}11) + I_m(111)\right) + I_t(101)} \qquad (2)$$

Here, I_t and I_m represent the integrated intensities of (111) tetragonal and of either (-111) or (111) monoclinic peaks, respectively.

Finally, Berkovich nanoindentation was performed with a XP nanoindenter equipped with a constant stiffness measurement modulus (CSM); this equipment was previously calibrated with fused silica. For each sample, 16 indentations of less than 2000 nm depth were performed and the hardness and Young modulus were automatically calculated as a function of depth. The methodology for obtaining these properties from the nanoindentation data is well-known and described elsewhere [12]. For each sample, the nanoindentation curves (for hardness and Young modulus) were averaged over all tests, and the standard deviation was calculated at each point of the final curves.

3. Results

The microstructure of the specimens is shown in figure 1. The standard 3Y-TZP (AS) had an average grain size of 0.30 μm; while the same ceramic produced by SPS had a much smaller grain size of 0.12 μm. On the other hand, the nanocomposite had a 70 % content of Ce-TZP and a 30 % content of Al_2O_3. The nanocomposites possessed the coarsest microstructure, with the Ce-TZP grains having an average grain size of 0.56 ± 0.20 μm, and it can be appreciated how they are larger than the alumina grains (which appear darker by SEM).

Figure 1. Scanning electron micrographs showing the microstructure of the three zirconia ceramics: (i) as-sintered 3Y-TZP (AS), (ii) Ce-TZP/Al_2O_3 nanocomposite and (iii) 3Y-TZP(SPS).

The nanoindentation load-displacement curves were very similar for the three ceramics (figure 2). This curve shows the real-time evolution of the applied indentation load as a function of the penetration depth. In all materials, the maximum indentation load applied was 635 mN, which produced a maximum displacement into the surface of 1480 nm for the AS and SPS materials. In the case of the Ce-TZP/Al_2O_3 nanocomposite, the maximum penetration depth for that load was slightly higher at 1510 nm. This is coherent with the slightly smaller hardness of this material already shown by the macroscopic Vickers indentation as compared to the AS ceramic (see table 1).

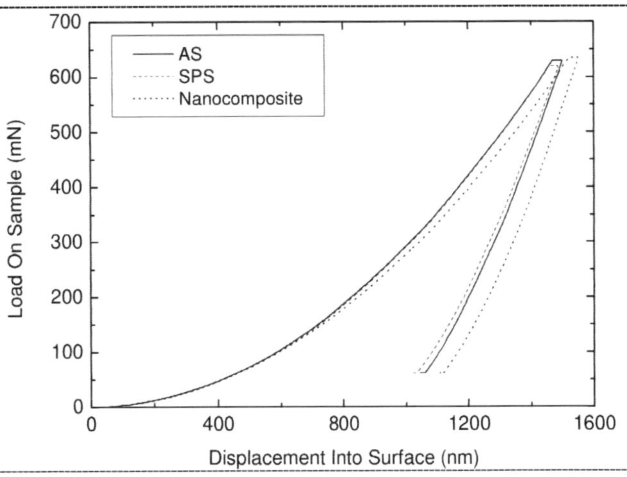

Figure 2. Nanoindentation load-displacement curves for the different samples: as-sintered 3Y-TZP (AS), 3Y-TZP(SPS) and Ce-TZP/Al_2O_3 nanocomposite.

Table 2. Young modulus and Berkovich hardness as averaged over the depth of the nanoindentation tests. These properties are shown for each material before and after being subjected to ageing. The monoclinic content formed at the surface during ageing is also shown.

Material	E (GPa)	E_{aged} (GPa)	H (GPa)	H_{aged} (GPa)	V_m
AS	250 ± 20	167 ± 44	17.5 ± 0.1	11.0 ± 2.0	70 %
SPS	230 ± 20	247 ± 20	17.5 ± 0.4	17.6 ± 0.3	< 5%
Ce-TZP/Al_2O_3	242 ± 77	263 ± 10	15.4 ± 0.6	15.6 ± 0.4	< 5%

The surface nanoindentation properties were plotted as a function of the penetration depth for each sample, before and after the ageing test (figures 3 and 4). Additionally, table 2 shows the average of these properties over the nanoindentation depth. For the AS material, the surface shows an initial Berkovich hardness of 17.5 GPa and a Young modulus of 250 GPa. After the ageing treatment, these surface properties decay substantially by more than 30 %. This is a consequence of the susceptibility of standard 3Y-TZP to ageing and is coherent with a previous nanoindentation work of Gaillard *et al* [9] in a very similar Y-TZP. In the fine-grained 3Y-TZP(SPS), the initial hardness and Young modulus at the surface are very similar to those of AS; however it can be clearly appreciated from the curves that both these surface properties are not affected by the ageing treatment. Effectively, both the hardness and Young modulus penetration curves remain effectively the same as can be appreciated in figures 3 and 4. Similarly, the surface Ce-TZP/Al_2O_3 nanocomposite maintained its mechanical properties at the surface as can be appreciated from the curves. It should be outlined that, for both the SPS and nanocomposite materials, the curves may seem only to be affected at small penetration depths in a distance of 100nm close to the surface; however even in this case as can be appreciated the error is too large to clearly conclude that ageing has affected these properties in this near-surface region. All these results are coherent with the amount of monoclinic content detected on the surface of the aged specimens (table 2): after ageing, V_m is 70 % for AS zirconia, while it was smaller than 5 % for the other two ceramics.

IOP Conf. Series: Materials Science and Engineering 31 (2012) 012015 doi:10.1088/1757-899X/31/1/012015

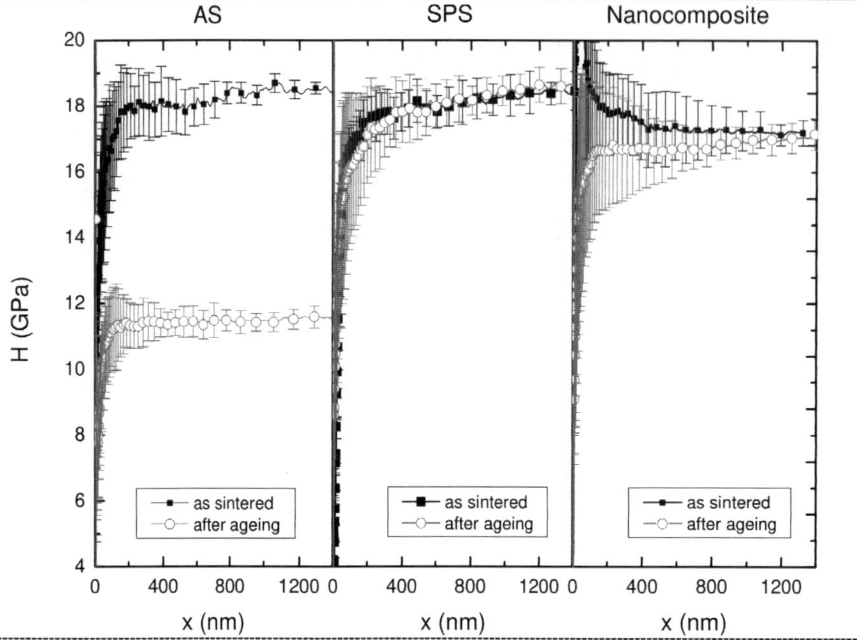

Figure 3. Hardness versus displacement into surface (x) for the three ceramics, before and after being subjected to ageing.

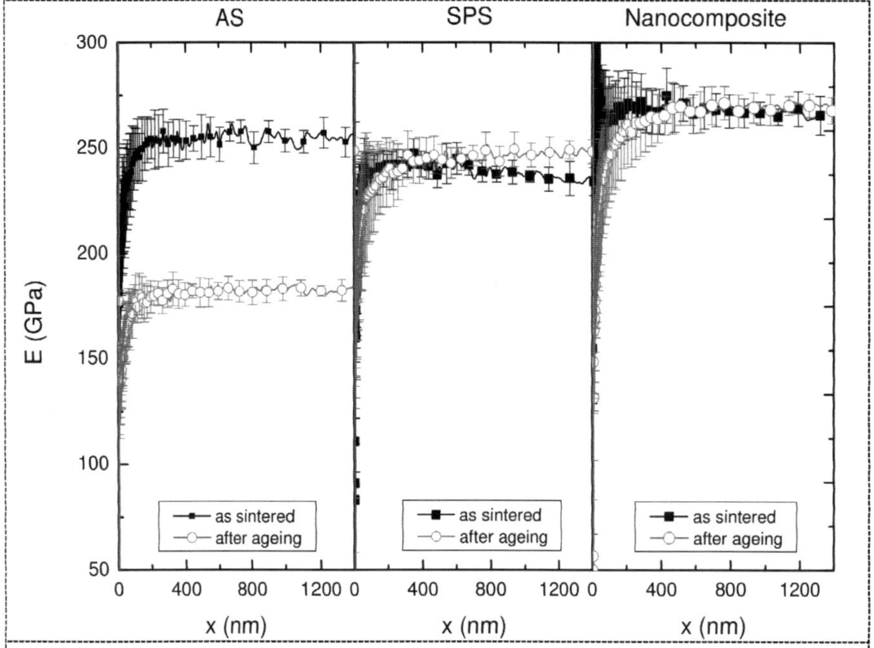

Figure 4. Young modulus versus displacement into surface (x) for the three ceramics, before and after being subjected to ageing.

4. Discussion

If the zirconia ceramic is susceptible to hydrothermal ageing, the surface suffers spontaneous t-m transformation to monoclinic phase and a layer is formed which has lower mechanical properties as compared to the healthy bulk. This layer presents microcracking which grows into the bulk with ageing time. Resistance to ageing can be achieved by avoiding the t-m transformation taking place in the presence of water. The driving force for this phase transformation depends on the grain size and type and amount of stabiliser used. Therefore in 3Y-TZP (SPS) the reduced susceptibility to ageing is achieved by the finer grain size achieved as compared to AS (see table 1). The reason is that, in SPS, sintering takes place at lower temperature (1200 °C in the present case) and in shorter times than in conventional sintering. There is a critical grain size below which no surface transformation will take place in the presence of water. For 3Y-TZP, this critical grain size is just below 0.30 μm. This explains why the present SPS material is not susceptible to ageing as it has a grain size of 0.12 μm. On the other hand, the Ce-TZP/ Al_2O_3 nanocomposite has a coarser microstructure with the zirconia grains having an average grain size of 0.56 μm, but despite this the zirconia component is resistant to ageing. In this case, the key factor enhancing the resistance to ageing is the choice of stabilizer in the tetragonal zirconia grains. TZP zirconia stabilized with CeO_2 increases its resistance to hydrothermal ageing. For example, commercially available 12Ce-TZP is known to show a much larger resistance to ageing than 3Y-TZP. Therefore zirconia doped with ceria in the nanocomposite is more resistance to ageing than 3Y-TZP with standard grain size.

4. Conclusions

The surface mechanical properties of 3Y-TZP sintered by SPS at 1200 °C and of a Ce-TZP/Al_2O_3 nanocomposite do not decay with hydrothermal exposure, as opposed to standard 3Y-TZP. Both the SPS-sintered 3Y-TZP and the Ce-TZP/Al_2O_3 nanocomposite maintain their surface integrity during exposure to 131°C hot water vapour in autoclave for 100 hours; which is an equivalent exposure to water as several years in vivo of an TZP implant. Thus, both those advanced ceramics can be regarded as potential candidates for biomedical applications.

5. References

[1] Kobayashi K, Kuwajima H and Masaki T 1981 *Solid State Ionics* vol 3-4 pp 489–93

[2] Lawson S 1995 *J. Eur. Ceram. Soc.* vol 15 pp 485-502

[3] Marro F G, Chintapalli R, Hvizdos P, Soldera F, Mucklich F, Anglada M, 2009 *International Journal of Materials Research* vol 100 pp 92-96

[4] Chevalier J, Gremillard L and Deville S 2007 *Annu. Rev. Mater. Res.* vol 37 pp 1–32

[5] Chevalier J, Gremillard L, 2009 *J. Eur. Ceram. Soc.* vol 29 pp 1245–55

[6] Nawa M, Bamba N, Sekino T and Niihara K 1998 *J. Eur. Ceram. Soc.* vol 18 pp 209–219

[7] Chintapalli R, Marro F G, Valle J A, Yan H, Reece M J and Anglada M 2009 *IOP Conf. Series: Materials Science and Engineering* vol 5 p 012014

[8] Marro F G, Mestra A and Anglada M 2011 *J. Eur. Ceram. Soc.* vol 31 pp 2189-2197

[9] Gaillard Y, Jiménez-Piqué E, Soldera F, Mücklich F and Anglada M 2008 *Acta Materialia* vol 56 pp 4206–4216

[10] Niihara K, Morena R and Hasselman D P H 1982 *J. Mater. Sci. Letters* vol 1 pp 13–16

[11] Toraya H, Yoshimura M and Somiya S 1984 *J. Am. Ceram. Soc.* vol 67 pp C119–121

[12] Oliver W C and Pharr G M 1992 *J. Mater. Res.* 7 vol pp 1564–1583

IOP Conf. Series: Materials Science and Engineering **31** (2012) 012016 doi:10.1088/1757-899X/31/1/012016

Development of a novel zirconia dental post resistant to hydrothermal degradation

E Camposilvan[1], F G Marro[1,2], A Mestra[1,2], M J Anglada[1]

[1]Department of Materials Science and Metallurgical Engineering, Universitat Politècnica de Catalunya, Av. Diagonal 647, 08028 Barcelona – Spain

[2]Center of Research in Nanoengineering CRnE, Universitat Politècnica de Catalunya

erik.camposilvan@upc.edu

Abstract. Tetragonal Zirconia Polycrystals stabilized with 3% mol. content of yttria (3Y-TZP) has excellent properties in terms of strength and fracture toughness. These properties are mostly imputable to the transformation toughening mechanism, by which the doped metastable tetragonal phase of zirconia transforms to monoclinic under applied stress ahead of a crack. This phenomenon is accompanied by a volume expansion of 5%, and increases the resistance to crack growth, thus leading to higher toughness and strength. An important drawback of this material is represented by the Low Temperature Degradation (LTD or aging), which consists in the progressive tetragonal-to-monoclinic phase transformation by the influence of water. This work focuses on the improvement of 3Y-TZP aging behavior in order to develop a novel dental post, by means of the addition of ceria from the surface. This was achieved through the impregnation of the pre-sintered samples with a solution containing Cerium, followed by sintering. Various pre-sintering temperatures were studied in terms of microstructure, mechanical properties and aging resistance. The novel zirconia dental posts developed in this work are much more resistant to LTD as compared to the base material with no loss in mechanical properties.

1. Introduction

Zirconia-based ceramics have been used in different engineering applications because of their excellent strength, fracture toughness, hardness and chemical stability. In the 90's, biomedical grade 3 mol% yttria-tetragonal polycrystalline zirconia (3Y-TZP) became an alternative to alumina as structural bio-inert ceramic in orthopaedics because of its substantially higher fracture toughness. The use of 3Y-TZP in orthopaedics paved the way to new implant designs that were not possible with alumina, which is more brittle. Examples of medical applications are smaller 3Y-TZP femoral heads, 3Y-TZP knees and, more recently, a large range of dental devices (crowns, bridges, abutments and dental implants) [1]-[3].

Pure zirconia exhibits three allotropes: monoclinic (*m*) phase from room temperature up to 1170 °C, tetragonal (*t*) phase from 1170 °C up to 2370 °C and cubic (*c*) phase for temperatures above 2370 °C. The *t* – *m* phase transformation is martensitic and is accompanied by a volume expansion of approximately 5%. In 3Y-TZP, yttria is employed to retain the *t* phase, in a metastable condition, down to room temperature, avoiding the *t* - *m* transformation and related expansion during cooling. In addition, stress can locally activate the transformation of the metastable phase; when this occurs ahead of a growing crack, the expansion produces a net compressive residual stress which contributes to hinder crack extension. The mechanism is called transformation toughening and results in a material with higher toughness [4].

The strong drawback of this material is represented by its poor resistance to Low Temperature Degradation (LTD or aging): the metastable *t* phase tends to convert spontaneously and progressively into the monoclinic one when the material is exposed to moderate temperatures and humid environment [5]. The transformation starts from the exterior and is accompanied by micro-cracking [6], leading to an increase in roughening and loss of surface mechanical properties [7], [8].

Although the kinetics of transformation is maximal at temperatures around 250 °C, there is now evidence that LTD is also possible even at human body temperature. This fact contributed to a high rate of failures for some batches of 3Y-TZP femoral heads in 2001, with the consequence that zirconia is not used anymore in these prosthesis [9].

Several authors have developed different methods for increasing the resistance to aging, as long as LTD sensitivity is affected by many variables such as grain size, stabilizer type and content, composition and homogeinity, residual stresses, temperature and environment, porosity and phase separation [7]. The use of stabilizers like Cerium oxide seems to be a promising way to reduce LTD [10]. The slower kinetics of degradation related with the use of Ce stabilizer appears to be due to the lower content of vacancies (Ce^{4+} does not generate vacancies in ZrO_2 lattice), which allows a reduced mobility of water species responsible for the LTD, if compared with the yttria stabilized material. Nevertheless, the addition of such stabilizers tends to compromise mechanical properties of 3Y-TZP, especially in terms of strength, as a result of grain growth [11], [12]. A way to improve aging resistance and maintain the bulk mechanical properties of the base material is represented by the diffusion of Cerium oxide from the surface after sintering [13].

In the common processing route for 3Y-TZP dental implants, sintering is split into two stages: pre-sintering and full sintering, the second at higher temperature. At the pre-sintered state, blanks are soft enough to allow machining via CAD-CAM technology in order to shape the final geometry. After machining, parts are sintered and post-processed.

In the present work, Cerium oxide is added through infiltration of a solution containing Cerium into the pre-sintered porous ceramic piece, previous to final sintering. In a preliminary study, the process was optimized investigating different ceria contents in terms of mechanical properties, chemical composition and aging resistance after sintering and accelerated aging. A set of dental implant prototypes was prepared according to the process previously developed and their flexural strength was measured before and after aging to verify that mechanical properties are not affected either by ceria introduction or by LTD. The final aim is to improve LTD resistance without affecting mechanical properties.

2. Materials and methods

The base material was 3Y-TZP powder provided by Tosoh Corp. In a preliminary study, samples were produced in form of discs, in order to facilitate their manipulation and test. Here, cylindrical prototypes approx. 16 mm long with 2.9 mm in diameter, i.e. with geometry similar to commercial dental implants, were manufactured.

The powder was compacted by cold isostatic pressing (CIP) at 200 MPa with a dwell time of 10 minutes. The rod compacted bodies were then pre-sintered in air during one hour in a tubular furnace (Hobersal ST-18). The heating and cooling rate was 3 °C/min. Half of the pre-sintered rods were submerged in a solution containing Cerium for 2 hours to ensure its infiltration in the open porosity. The prototypes were then sintered at 1450 °C during 2 hours with a heating/cooling rate of 3 °C/min and tested "as fired".

2.1. *Aging sensitivity*

The degradation process was simulated in autoclave under water vapour at 131 °C, 2 bar, during 30 h, for the doped and undoped prototypes. According to the degradation kinetics studied by other authors

[14], this treatment corresponds roughly to more than an average human lifetime at the body temperature (37 °C) and atmospheric pressure. Thus, the process may be used as a good indicator of resistance to LTD for the lifespan of an implant. To quantify aging resistance, monoclinic phase content was measured after artificial aging through X-ray diffraction (XRD) analysis on mirror-polished samples. The equipment was a Bruker AXS D8 Advance diffractometer using Cu-Kα radiation and θ/2θ configuration. The monoclinic phase was calculated employing the equation proposed by Toraya *et al.* [15].

2.2. Mechanical characterization

Density and apparent porosity were measured using the Archimedes' method in distilled water, after pre-sintering and full sintering. Mechanical tests were carried on before and after artificial aging. Hardness and fracture toughness were determined by the Vickers indentation method, applied on the polished surface of the samples. For fracture toughness, the formula of Niihara et al. was employed [15]. Three-point bend test was performed on prototypes to evaluate their flexural strength, before and after aging treatment, for doped and undoped conditions (base material). Crosshead speed was set at 0.001 mm/s as indicated by the standard ASTM C 1161-02c [17].

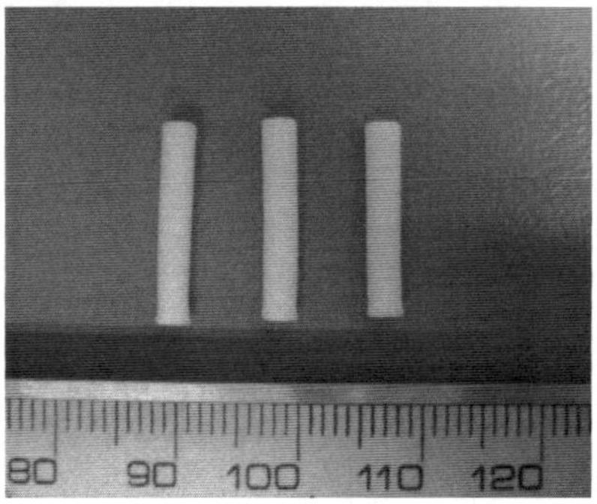

Figure 1. Prototypes after sintering.

2.3. Microstructural and elemental characterization

The microstructure was evaluated from SEM micrographs of mirror-polished and thermally etched samples. The average grain size was calculated using commercial image processing software. The chemical composition of the material was determined using a CAMECA SX50 Electron Probe and the results were converted to a molar basis (CeO_2 mol%).

3. Results and discussion

The measured density at the pre-sintered state was of 3.17 ± 0.05 gr/cm^3, 51.7 % of the theoretical one. An apparent porosity of 47.2 ± 0.2 % was calculated. After sintering, the density was 6.08 ± 0.02 gr/cm^3, reaching almost the theoretical value.

Electron Probe Micro Analysis (EPMA) measured CeO_2 surface contents of approx. 4 mol%. Thus, significant amounts of ceria were introduced into the material. Moreover, repeating the measurement along the cross section of the specimens, it was shown that less ceria reached the bulk, leading to a value of approx. 2.5 mol% at 1 mm depth from the surface.

The grain size calculated from analysis of SEM micrographs was similar for the doped and undoped specimens, with values of the average circular diameter around 0.35 μm, as showed in Figures 2 and 3. The hardness resulting from Vickers testing (load = 1 and 30 Kg) was of 12.8 ± 0.5 GPa for specimens either with and without ceria, before and after aging treatment. Indentation fracture toughness tests gave values of approx. 4.5 MPa·√m for all the conditions. Therefore, it can be stated that ceria addition did not modify substantially the microstructure of the material and its mechanical properties measured on the surface.

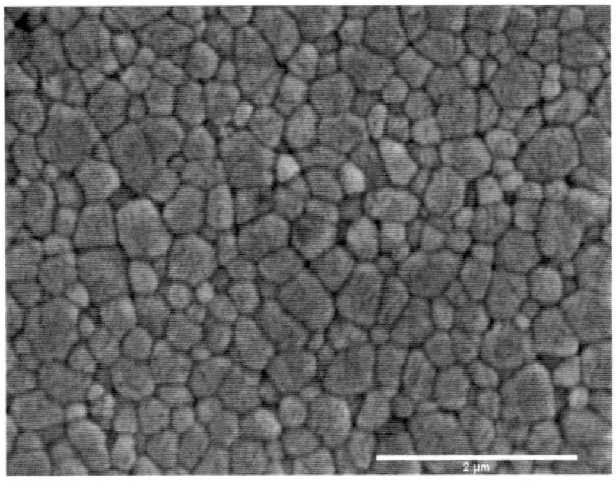

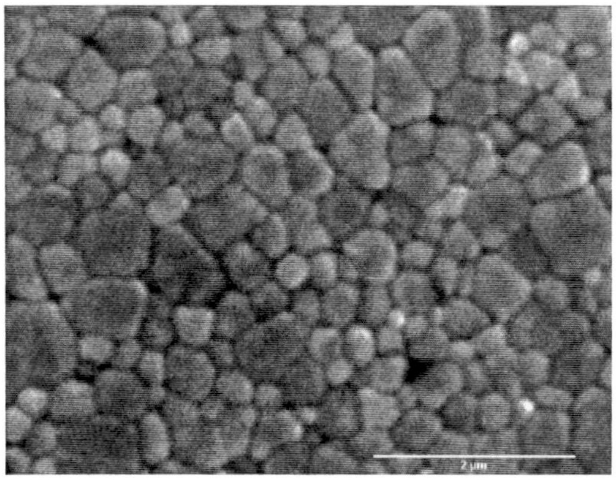

Figures 2, 3. SEM micrographs of the base material (left) and the ceria doped one (right). Scale = 2 µm.

As shown in Figure 4, the diffraction patterns of the base and doped materials were quite different after aging. For the base material, the *m* phase peaks were noticeable, resulting in a calculated 60% approx. of transformed phase. On the other side, for the doped material, *m* peaks were barely perceptible, corresponding to only approx. 5% of *m* phase. The treatment was then effective in improving the aging resistance.

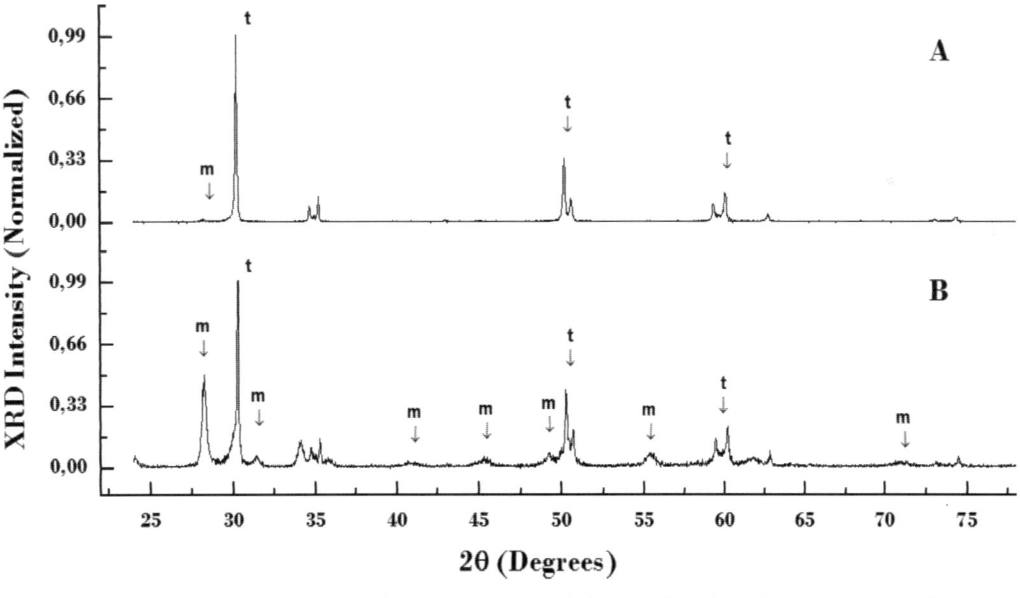

Figure 4. XRD patterns of aged specimens. (A) Material doped with ceria, (B) 3Y-TZP base material. The representative peaks for monoclinic (m) and tetragonal (t) phases are highlighted as they appear on the spectra.

The results of the three-point bend test are presented in Figure 5. The average strength was ranging between 530 and 600 MPa for all the four conditions of the material tested. Therefore, there was no evidence that the treatment proposed for ceria addition could affect the mechanical strength of the prototypes. These values do not represent the true mechanical strength of the material as long as the surface of the prototypes had not been prepared, in terms of defects elimination, before testing. Nevertheless, they provide a way to compare the strength of prototypes processed under different conditions and the effect of ceria on actual devices.

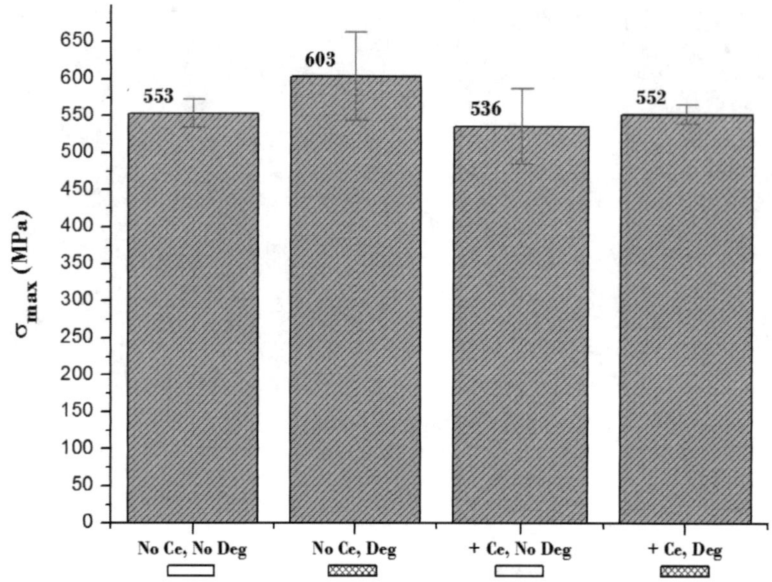

Figure 5. Three-point bend test results. Legend: white color, only 3Y-TZP; light-yellow colour, Ce doped-3Y-TZP; blank, tested "as sintered"; crosshatched, tested after aging. Red bars represent the standard deviations.

4. Conclusions

Ceria was effectively introduced into the material using the impregnation method. The amount of ceria was higher in the superficial regions than into the bulk, reaching a maximum of 4 mol%. The microstructure of the material was not appreciably modified by the infiltration. Mechanical properties were no significantly affected by ceria infiltration, both before and after artificial aging, satisfying the requirements for biomedical 3Y-TZP. The introduction of ceria into the ceramic was effective in reducing LTD after artificial aging. The results of three-point bend test on the implant prototypes revealed that the treatment proposed does not affect the strength of such devices.

Acknowledgments

This work was supported by TECNIO (Generalitat de Catalunya) under the project ref. VALTEC 08-1-0030. E Camposilvan acknowledges the Universitat Politècnica de Catalunya for the scholarship FPU-UPC.

References

[1] Christel P 1989 Zirconia: the second generation of ceramics for total hip replacement *Bull. Hosp. Joint. Dis. Orthop. Inst.* **49** 170–177.

[2] Piconi C and Macauro G 1999 Zirconia as a ceramic biomaterial *Biomat.* **20** 1–25.

[3] Denry I and Kelly R 2008 State of the art of zirconia for dental applications *Dent. Mater.* **24** 299-307.

[4] Garvie R C, Hannink R H and Pascoe R T 1975 Ceramic steel? *Nature* **258** 703–704.

[5] Lawson S 1995 Environmental degradation of zirconia ceramics *J. Eur. Cer. Soc.* **15** 485-502.

[6] Kobayashi K, Kuwajima H and Masaki T 1980 Phase change and mechanical properties of ZrO_2 – Y_2O_3 solid electrolyte after ageing *Solid State Ion.* **3/4** 489-93.

[7] Chevalier J and Gremillard L 2009 The tetragonal-monoclinic transformation in zirconia: lesson learned and future trends *J. Am. Ceram. Soc.* **92** 9 1901-1920.

[8] Guo X 2004 Property Degradation of Tetragonal Zirconia Induced by Low-Temperature Defect Reaction with Water Molecules *Chem. Mater.*, **16** 21 3988–3994.

[9] Chevalier J *et al.* 2011 Reliability assessment in advanced nanocomposite materials for

orthopaedic applications, *J. Mech. Behav. of Biom. Mat.* **4** 3 303-314.

[10] Lin J D and Duh J G 2002 Mechanical properties and resistance to hydrothermal aging of ceria- and yttria-doped tetragonal zirconia ceramics *Mat. Chem. Phys.* **77** 808-818.

[11] Tsukuma K and Shimada M 1985 Strength, Fracture Toughness and Vickers Hardness of CeO_2-Stabilized Tetragonal ZrO_2 Polycrystals (Ce-TZP) *J. Mat. Sci.* **20** 1178-1184.

[12] Lin J D and Duh J G 2002 Correlation of Mechanical Properties and Composition in Tetragonal CeO_2-Y_2O_3-ZrO_2 Ceramic System *Mat. Chem. Phys.* **78** 246-252.

[13] Marro F G, Valle J, Mestra A and Anglada M 2011 Surface modification of 3Y-TZP with Cerium oxide, *J. Eur. Ceram. Soc.* **31** 3 331–338.

[14] Chevalier J, Drouin J M and Cales B 1997 Low temperature aging of zirconia hip joint heads *Proceedings of the 10th International Symposium on Ceramics in Medicine* ed L Sedel, C Rey (Oxford, UK: Elsevier Science Ltd) 135–138.

[15] Toraya H, Yoshimura M and Somiya S 1984 Calibration curve for quantitative analysis of the monoclinic-tetragonal ZrO_2 system by X-Ray diffraction *J. Am. Ceram. Soc.* **67** 6 C-119–C-121.

[16] Niihara K, Morena R and Hasselman D P H 1982 Evaluation of K_{Ic} of brittle solids by the indentation method with low crack-to-indent ratios *J. Mater. Sci. Letts.* **1** 13-16.

[17] ASTM C 1161-02c 2002 Standard test method for flexural strength of advanced ceramics at ambient temperature.

Thermal treatment and phase formation in kaolinite and illite based clays from tropical regions of Bolivia

E Escalera[1], M L. Antti[1] and M Odén[2]
[1]Division of Materials Science, Luleå University of Technology, 97187 Luleå, Sweden
[2]Nanostructured Materials, Department of Physics, Chemistry and Biology, Linköping University, 581 83 Linköping, Sweden

E-mail: edwin.escalera@ltu.se

Abstract. The aim of this study was to compare the thermal behaviour of clays containing illite and kaolinite in various proportions. The clays contained small amounts of K and Fe, which act as fluxing agents. In order to investigate the phase formations during heating, the samples were examined in a differential scanning calorimeter at temperatures up to 1300°C. The thermal expansion of the samples was determined by dilatometer measurements from room temperature up to 1150°C. Phases were identified using x-ray diffraction and scanning electron microscopy.

In all samples, most of the kaolinite was transformed into metakaolinite during heating up to 650°C, while illite remained unchanged up to 950°C. There was no influence of K and Fe on dehydroxylation. Metakaolinite formed at temperatures above 950°C leading to a Si-Al spinel. Furthermore, mullite was formed in the temperature interval 1050-1150°C. In this temperature range, the mechanism of mullite formation depended on the amount of K and Fe in the samples, changing the temperature of formation of mullite. It was observed by x-ray diffraction that most of the illite was transformed into a Si-Al spinel phase at 1050°C, and during further heating transformed into mullite. An increased amount of illite in the clays slightly decreased the melting temperature. The dilatometer measurements showed expansion and shrinkage for the dehydroxylation and spinel-phase formation, respectively.

1. Introduction

Clay materials from tropical regions of Bolivia, located in the eastern lowlands including large sections of Amazonian rainforest and the Chaco plain, are promising materials for applications such as whitewares, tiles, filters and bricks. The suitability of the clay in these applications is strongly dependent on its structure, composition and physical properties.[1]

The most common commercial clay minerals for traditional applications are kaolinite, illite and montmorillonite.[2, 3] Feldspars, along with many other aluminosilicate minerals, have transformed to different types of clays over long periods of time by gradual chemical weathering.[4] The most common secondary components in natural clays are iron hydroxides, quartz, feldspars, pyrite, carbonates and other insoluble salts.[5] These components largely affect clay characteristics and may influence its use in specific applications.

Bolivia has a variety of geographical zones rich in non-metallic minerals such as calcite, dolomite, talc, feldspar, clays, gypsum, ulexite and others.[6]

[1] Corresponding author: edwin.escalera@ltu.se

The aim of this study was to characterize two types of clays found in the eastern tropical region of Bolivia, in terms of their mineralogical and chemical composition, using their thermogravimetric behaviour with respect to phase formation and microstructural evolution. These clays are candidates as raw materials for different types of building materials.

2. Materials and methods

The Chapare region hosts the largest deposits of clays found in Bolivia. Ivirgarzama clay (IC) and Entre Rios clay (EC) deposits are located 220 and 290 km, respectively, from Cochabamba Department. The coordinates and locations of the two main deposits are given in Figure 1. Samples from these two deposits were collected by extraction of larger amount of clays (20 kg) to ensure representative samples.

Initially, the samples were dried at 110ºC and desegregated manually. Subsequently, 100 g of each sample was ground to fine powder in an agate mortar, and screened through a 315 mesh size sieve (DIN 0.315). These samples have been used for all experiments.

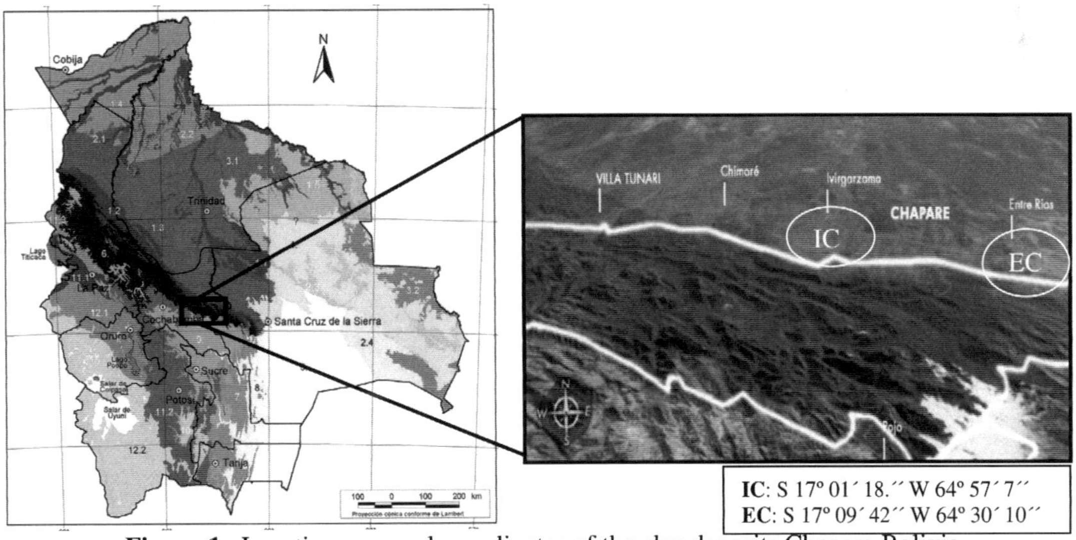

Figure 1. Location map and coordinates of the clay deposits Chapare-Bolivia.

2.1. Mineralogical and chemical analyses

The mineralogical analyses of the raw clay samples, as well as analyses of their corresponding phase composition after firing, were carried out by x-ray diffraction (XRD) techniques. The XRD patterns were obtained using a Siemens D 5000 X-ray diffractometer with CuKα_1 radiation (λ = 1.5418 Å) operating at a tube voltage and current of 40 kV and 20 mA, respectively. Diffraction patterns were recorded in the range between 5 and 45° (2θ), and were compared with existing patterns in powder diffraction files (PDF).

The clays were fired in a chamber furnace in an ambient atmosphere at different temperatures: 450, 650, 950, 1050, 1150 and 1250ºC. The heating rate was 10°C/min. Samples were subsequently furnace cooled. The temperatures were chosen from the results of the thermal analysis as described below.

Chemical analyses of as-extracted clays were performed using Inductively Coupled Plasma Atomic-Emission Spectroscopy (ICP-AES).

2.2. Thermal analysis

Differential scanning calorimetry / thermal gravimetry (DSC/TG) was performed to investigate the phase transformations and mass changes during heating. DSC/TG experiments were performed in a Netzsch STA 449C Jupiter instrument, equipped with a Netzsch Aeolos QMS 403C mass spectrometer. The experiments were carried out on powder samples in an alumina crucible between room temperature and 1300°C at a heating rate of 10°Cmin^{-1} in an air atmosphere.

The dilatometry experiments were performed in a Netzsch DIL 402C instrument. Prior to measurement, the compacted sample was dried overnight at 110°C. The samples were heated from room temperature and 1150°C in an air atmosphere at a heating rate of 10°Cmin^{-1}. The density and open porosity of samples heat treated at 1150°C in the dilatometer were determined using Archimedes principle.

The morphology of the as-extracted clay powders was studied by scanning electron microscopy (SEM) using a JEOL 6460-LV microscope. The plasticity index of both samples in their raw states was obtained through determination of the Atterberg limits. These are defined as the liquid limit (LL) and plastic limit (PL), according to the ASTM standard test method D4318.

3. Results and discussion

3.1. Mineralogical analysis

XRD curves from the two types of clays investigated (IC and EC) are shown in Figure 2. Both clays contain kaolinite and illite, belonging to the main group of phyllosilicates. The peaks corresponding to quartz have a relatively high intensity in the IC clay compared with the EC clay, which indicates that a larger amount of free silica exists in the IC clay.

XRD analyses reveal that the EC clay contains montmorillonite along with illite corresponding to smectite and mica structures, respectively. The mineral formulae for kaolin structures and mica structures are written $Al_2O_3 \cdot 2SiO_2 \cdot 2H_2O$ and $K_2O \cdot 3Al_2O_3 \cdot 6SiO_2 \cdot 2H_2O$, respectively.[7]

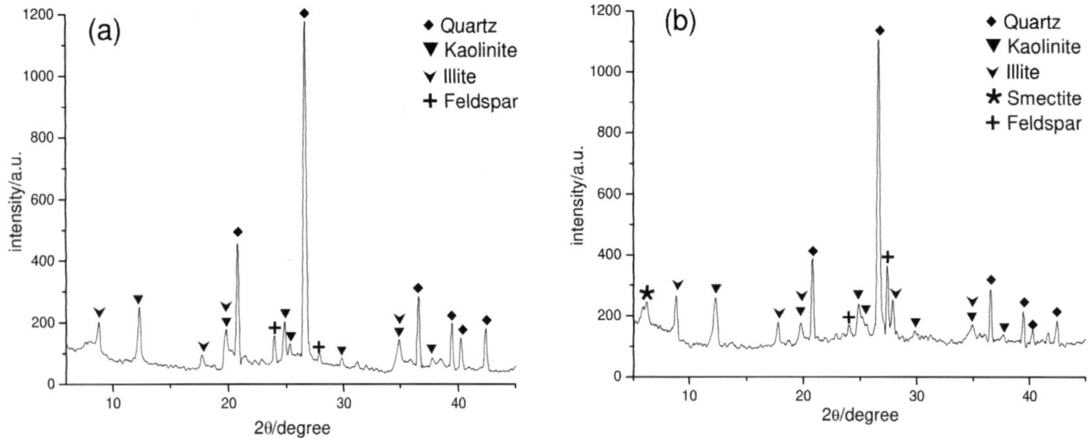

Figure 2. X-ray diffraction patterns for (a) IC clay and (b) EC clay.

Table 1 shows the chemical composition of the clays. The IC clay had low levels of iron and potassium as fluxing agents, in comparison with the EC clay. An iron content of 6 wt% in the EC clay was sufficient to develop a reddish colour after firing. Iron oxide (Fe_2O_3) is known to be the main colorant in clays after firing.[8, 9] Both clays show SiO_2 and Al_2O_3 to be the predominant oxides. The absence of a correlation between Al_2O_3 and SiO_2 contents indicates that excess SiO_2 is a result of the presence of free quartz, as shown also by XRD (Figure 2). The higher SiO_2/Al_2O_3 ratio corresponds to the IC clay (4.1) compared with the EC clay (3.5) according to the results in table 1. These ratios are replicated through the intensity of the peaks by XRD, indicating a larger amount of free silica in the IC clay.

Table 1. Chemical composition (wt%) of IC and EC clays.

	SiO_2	Al_2O_3	Fe_2O_3	K_2O	Na_2O	CaO	MgO	TiO_2	LOI*
IC clay	71.8	17.5	1.67	2.32	0.12	0.10	0.81	0.96	6.10
EC clay	65.9	18.8	5.87	3.39	0.82	0.18	1.76	0.87	6.75

*LOI = loss on ignition

The incorporation of iron and potassium in clays is likely to occur through an isomorphous substitution of Fe^{3+} and K^+ cations in the clay mineral structures on octahedral sites.[10] This normally occurs in mica structures, resulting in higher interlayer spacings in mica compared with kaolinite structures.[11] The relatively high amounts of Fe_2O_3 and K_2O oxides shown in table 1 are a result of the presence of illite and montmorillonite, mainly in the EC clay, as detected by XRD.

Another important aspect regarding the chemical composition of both clays is the presence of alkali components such as K_2O and Na_2O, which act as fluxing agents, i.e. they decrease the firing temperature.

The microstructural development of fired clays has been the subject of several recent studies. Brindley et al.[12] suggested that the transformation from kaolinite to mullite can be described by the following reactions, which occur during heating to 1150ºC:

$$\underset{\text{kaolinite}}{Al_2O_3\cdot 2SiO_2\cdot 2H_2O} \rightarrow \underset{\text{metakaolinite}}{Al_2O_3\cdot 2SiO_2} + H_2O \quad (1)$$

$$\underset{\text{Metakaolinite}}{Al_2O_3\cdot 2SiO_2} \rightarrow \underset{\text{spinel}}{2Al_2O_3\cdot 3SiO_2} + \underset{\text{amorphous}}{SiO_2} \quad (2)$$

$$\underset{\text{spinel}}{2Al_2O_3\cdot 3SiO_2} \rightarrow \underset{\text{mullite}}{3Al_2O_3\cdot 2SiO_2} + \underset{\text{amorphous}}{SiO_2} \quad (3)$$

XRD patterns of fired clay samples are shown in Figure 3. They show that kaolinite is completely transformed into metakaolinite at 650ºC in both clays since all reflections have disappeared. Metakaolinite is a result of removing the hydroxyl groups above 400°C, as shown in equation (1) and is well known as a highly disordered phase.[13, 14] Because of this disorder it could not be identified by XRD.

Illite peaks are present until the temperature has reached 950ºC, as observed in Figure 3 for both clays, i.e. the original crystal structure is maintained at low amounts at this temperature (950°C). These illite peaks have disappeared completely at 1050°C. This may be caused by the illite structure breaking down in this range of temperature. Hematite begins to crystallize at 1050°C and remains present until 1250°C, see Figure 3b.

Mullite is detected in both samples at 1150°C through the appearance of the characteristic XRD peaks and its amount increases at 1250°C (Figure 3). The mullite peaks are higher in the EC clay than in the IC clay, this may be a result of excess silica being liberated as amorphous silica. At this temperature, mullite and quartz are the only phases remaining in both clays.

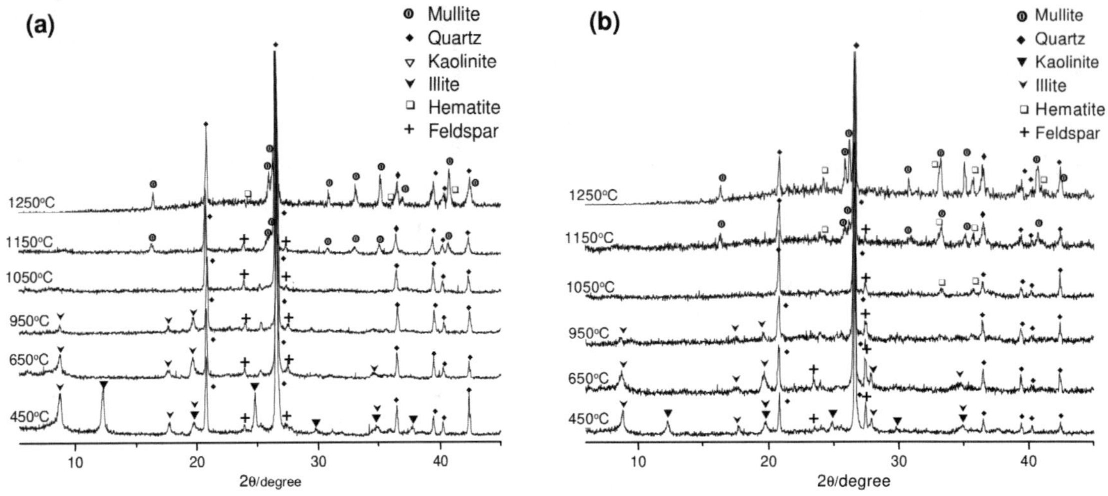

Figure 3. X-ray diffraction patterns of fired clays at different temperatures for (a) IC clay and (b) EC clay.

3.2. Thermal analysis

The results obtained by DSC are shown in Figure 4. Phase changes in the clays studied are indicated by endothermic and exothermic peaks; the temperatures corresponding to these peaks are given in table 2. Mass losses as well as suggested types of reactions are also given in the table.

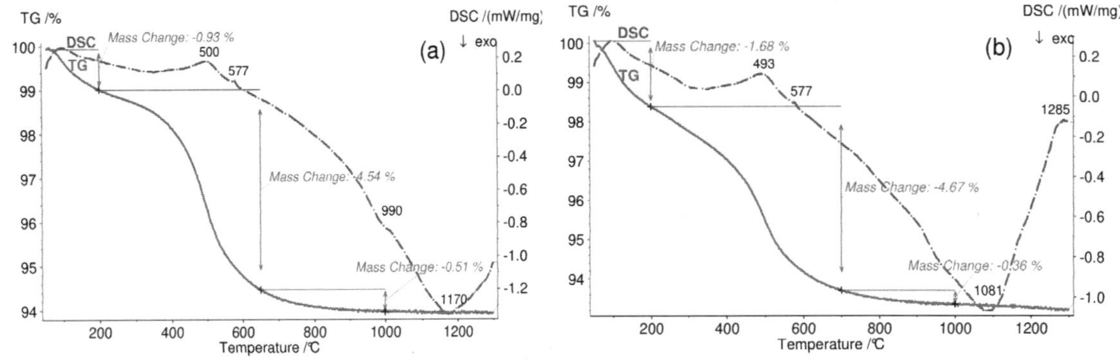

Figure 4. DSC and TG curves for (a) IC clay and (b) EC clay.

Table 2. Differential thermal analysis data for the IC and EC clays.

Sample	**Mass loss (wt%) (Temp. range 25-1000°C)**	**Peak at max. temperature (°C)**			
		Dehydroxylation	α-β quartz inversion	Si-Al spinel	Mullite phase formation
IC clay	5.98	500	577	990	1170
EC clay	6.71	493	577	---	1081

The DSC curve for the IC clay (Figure 4a) shows an endothermic peak at 500°C, which results from the dehydroxylation of kaolinite, followed by another endothermic peak at 577°C which is a result of the α↔β quartz inversion. An exothermic peak at about 990°C associated with the formation of Si-Al spinel according to equation (2) is also seen. The total mass loss for the IC clay is 5.98% from room temperature to 1000°C, associated with the liberation of water absorbed at 100°C and hydroxyl groups (OH) from the lattice structure of kaolinite and illite.[15] The rather abrupt change in mass loss rate for the IC clay (Figure 4a) starting at around 350ºC and continuing until 650°C is an indication of dehydroxylation according to equation (1).

The DSC curve of the EC clay (Figure 4b) shows an endothermic peak at 493°C resulting from the dehydration of kaolinite. Illite and montmorillonite dehydroxylate below 600°C and form a stable dehydroxylated phase, which correlates with the broadening of the XRD peaks seen in Figure 3b.The total mass loss associated with these effects is 6.71% for the EC clay (see table 2).

Predominant exothermic peaks are seen at 1170°C for the IC clay and at 1081°C for the EC clay, which are related to mullite formation according to equation (3). The existence of the mullite phase is supported by the XRD results of the clays fired at these temperatures (Figure 3).

The dilatometer curves for both clays are shown in Figure 5. From room temperature to 450°C, regular expansion is observed. From 450 to 550°C, the slight shrinkage observed is related to dehydroxylation of kaolinite and its transformation into metakaolinite. Both clays exhibit expansion at 577°C resulting from the α↔β quartz inversion. However, expansion is more pronounced in the IC clay. (Figure 5a)

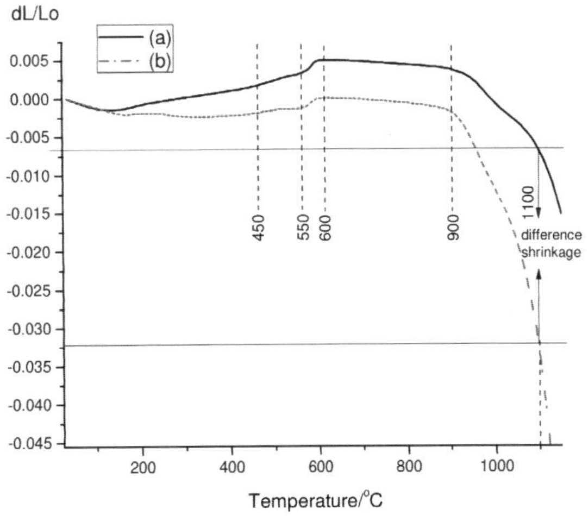

Figure 5. Dilatometer measurements of (a) IC clay and (b) EC clay.

The shrinkage between 600 and 850°C observed in both samples is believed to be caused by structural reorganization of metakaolinite into the Si-Al spinel phase. The pronounced shrinkage in the range 900 to 1150°C during the dilatometer runs indicates the start of sintering of the clays. A remarkable difference in shrinkage between IC clay and the EC clay at 1100°C can be seen in figure 5, which indicates that the EC clay is able to form more amorphous silica (glassy phase) at lower temperature in comparison with IC clay.

The bulk density of the IC and EC clays after dilatometer runs up to 1150ºC were 2.10 and 2.47 gcm^{-3}, respectively. The open porosity was 17,2% for the IC clay and 1,2% for the EC clay. This indicates that the EC clay achieved a higher densification than the IC clay at this temperature.

IOP Conf. Series: Materials Science and Engineering **31** (2012) 012017 doi:10.1088/1757-899X/31/1/012017

3.3. Scanning electron microscopy

The SEM micrographs (Figure 6) of the raw powder samples show the morphology of the kaolin particles. The kaolinites are irregular plate-like sheets. Figure 6(b) shows some large kaolin flakes stacked together with montmorillonite and illite forming agglomerates as detected by XRD (Figure 2b). The agglomerates can be explained by the fact that montmorillonite absorbs water from the ambient air.

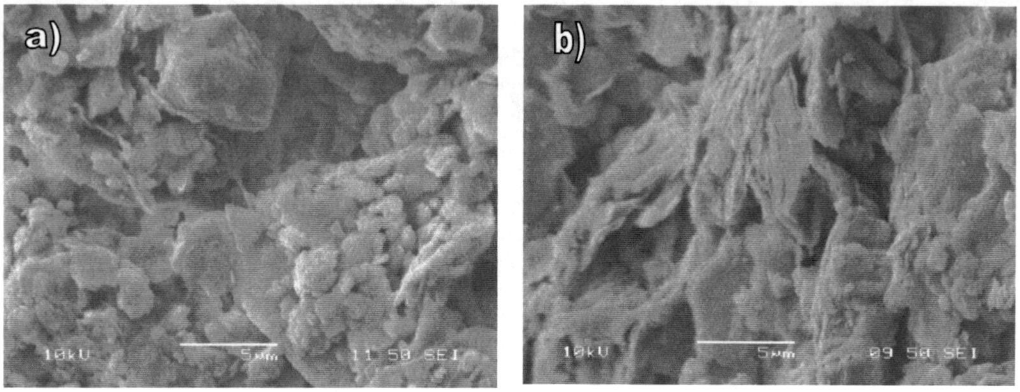

Figure 6. SEM micrographs showing the morphology of (a) IC clay and (b) EC clay.

3.4. Plasticity

The plastic limit and the plasticity index for IC clay were 20.50 and 21.27%, respectively. For the EC clay they were 24.55 and 26.72%, respectively. These high values of plasticity are typical for clay materials and indicate that the samples in this study have good plasticity according to the values defined by the ASTM standard test method D4318.

4. Conclusions

The characteristics of two types of clay from Chapare region of Bolivia and the effect of firing temperatures on phase obtained lead to the following conclusions:

- Kaolinite and illite are present in both clays together with free quartz.
- Montmorillonite and iron hydroxide were present in the EC clays, but not in the IC clays.
- The kaolinite in the clays transforms into mullite through several steps in the temperature interval 1050-1150°C during heating.
- In both samples, most of the kaolinite dehydroxylated during heating up to 650°C, while illite remained unchanged up to 950°C.
- The K and Fe content had no influence on dehydroxylation, but slightly influenced the mullite phase transformation.
- Most of the illite present in both samples was transformed into a Si-Al spinel phase up to 1050°C.
- The major amount of illite in the EC clay favours the sintering step permitting the formation of liquid phase at lower temperature.

From the results obtained in this study, the clays from IC and EC deposits are considered to be promising materials for ceramics manufacture, and will be investigated further based on their mechanical properties.

IOP Conf. Series: Materials Science and Engineering **31** (2012) 012017 doi:10.1088/1757-899X/31/1/012017

4. Acknowledgments

This work was financially supported by the Swedish International Development Cooperation Agency, SIDA. The authors wish to thank Dr John Ion for his help with the English revision.

5. References

[1] Ferrari S, Gualtieri A 2006 *Appl. Clay Sci.* **32** 73-81
[2] Siddiqui M, Ahmed Z, Saleemi A 2005 *Appl. Clay Sci.* **29** 55-72
[3] Celik H 2010 *Appl. Clay Sci.* **50** 245-54
[4] Shackelford F, Doremus H 2008 *Springer* 116-19
[5] Soro N, Aldon L, Oliver J, Jumas C, Laval P, Blanchart P 2003 *J. Am. Ceram. Soc.* **86** 129-34
[6] Montes de Oca I 1997 *Edobol* 608-14
[7] McConville C, Lee 2005 *J. Am. Ceram. Soc.* **88** 2267-76
[8] Stepkowska E, Jefferis S 1992 *Appl. Clay Sci.* **6** 319-42
[9] Dionicio A, Braga M, Waerenborgh J 2009 *Appl. Clay Sci.* **42** 629-38
[10] Manhaes R, Auler L, Sthel M, Alexandre J, Massunaga M, Carrio J, dos Santos D, da Silva E, Gracia-Quiroz A, Vargas H 2002 *Appl. Clay Sci.* **21** 303-11
[11] Konan K, Peyratout C, Bonnet J, Smith A, Jacquet A, Magnoux P, Ayrault P 2007 *J. Colloid Interface Sci.* **307** 101-08
[12] Brindley G, Nakahira M 1959 *J. Am. Ceram. Soc.* **42** 311-14
[13] Castelein O, Soulestin B, Bonnet J, Blanchart P 2001 *Ceram. Int.* **27** 517-22
[14] Chen C, Lan G, Tuan W 2000 *Ceram. Int.* **26** 715-20
[15] Lee W.E, Souza G.P, McConville C, Tarvornpanich T, Iqbal Y 2008 *J. Eur. Ceram. Soc.* **28** 465-71

Advances in Computational High-Resolution Mechanical Spectroscopy HRMS. Part I - Logarithmic Decrement

M Majewski[1], A Piłat[2], L B Magalas[1]

[1]AGH University of Science and Technology, Faculty of Metals Engineering and Industrial Computer Science, al. Mickiewicza 30, 30-059 Kraków, Poland
[2]AGH University of Science and Technology, Faculty of Electrical Engineering, Automatics, Computer Science and Electronics, al. Mickiewicza 30, 30-059 Kraków, Poland

E-mail: magalas@agh.edu.pl

Abstract. The comparison between different methods used to compute the logarithmic decrement in high-resolution mechanical spectroscopy (HRMS) is analyzed. The performance of parametric OMI method (*Optimization in Multiple Intervals*) and interpolated discrete Fourier transform (IpDFT) methods are investigated as a function of the sampling frequency used to digitize free decaying oscillations in low-frequency resonant mechanical spectrometers. It is clearly demonstrated that a new Yoshida-Magalas (YM) method is the most powerful IpDFT-based method which outperforms the standard Yoshida (Y) method and other DFT-based methods. Four IpDFT methods and the OMI method are carefully analyzed as a function of the sampling frequency. The results presented in this work clearly show that the relative error in the estimation of the logarithmic decrement depends both on the length of free decaying signal and on the sampling frequency. The effect of the sampling frequency was not yet reported in the literature. The performance of different methods used in the computations of the logarithmic decrement can be listed in the following order: (1) the OMI, (2) the Yoshida-Magalas YM, (3) the Yoshida-Magalas YM_C, and finally (4) the Yoshida Y.

1. Introduction

High-resolution mechanical spectroscopy HRMS [1-3] requires new computing tools and algorithms to determine the logarithmic decrement δ and the resonant frequency f_0 from free decaying oscillations with high precision and very low dispersion in experimental points. In addition, it is also expected that computations of the δ and f_0 will be independent on small external perturbations (i.e. the ZPD effect [7, 9], defined as the deployment of the center of damped harmonic oscillations) while computing time must be reasonably short. These, frequently contradictory requirements, are difficult to be fulfilled [1-20]. These problems are tackled in this work. Computations of the logarithmic decrement δ will be analyzed here as a function of the length of free decaying oscillations for two different sampling frequencies: 1 kHz (usually used in low-frequency resonant mechanical spectrometers) and 6 kHz (it will be demonstrated here that f_S = 6 kHz yields the best results for low-frequency mechanical spectrometers operating around the resonant frequency $f_0 \approx 1$ Hz). It should be emphasized that computation techniques used in HRMS depend on too many parameters, which is why the computational problem is a multi-dimensional task [1-13].

To elucidate this problem the effect of the length of the signal L and the sampling frequency f_S on computed values of the logarithmic decrement are investigated.

HRMS can provide better insight into a number of relaxation processes involving e.g. interaction of dislocations with mobile points defects (Dislocation-Enhanced Snoek Effect DESE and Snoek-Köster relaxation in bcc metals and alloys [14-20]), Bordoni relaxations, study of phase transitions and a number of transient phenomena frequently observed in mechanical loss measurements of different materials. It is noted that the OMI method (*Optimization in Multiple Intervals*) [1-8] and the Yoshida-Magalas YM method can be successfully used to analyze free decaying oscillations biased by the ZPD effect [7, 9] which accompanies phase transitions and dislocation-induced phenomena.

2. Results and Discussion

The exponentially damped time-invariant harmonic oscillations (free decaying oscillations, $A(t)$) embedded in an experimental noise $\varepsilon_w(t)$ can be described using the digitized data $A_i(t)$ and t_i acquired from free decaying signal [1, 2]:

$$A(t) = A_0 e^{-\delta f_0 t}\cos(2\pi f_0 t + \varphi) + \varepsilon_w(t) + dc\,, \tag{1}$$

where A_0 is the maximal strain amplitude of a sample mounted in a mechanical spectrometer, t is a continuous time in seconds, $-\pi < \varphi \leq \pi$ is the phase of the signal $A(t)$ in radians, and dc is an offset. The noise $\varepsilon_w(t)$ corresponds here to the signal-to-noise ratio S/N= 32 dB [1-7].

The logarithmic decrement δ can be computed from Eq. (2) [12]

$$\delta = 2\pi \frac{Im\left(\dfrac{-3}{R-1}\right)}{Re\left(s_1 - \dfrac{3}{R-1}\right)} \tag{2}$$

while

$$R = \frac{F(s_1) - 2F(s_2) + F(s_3)}{F(s_2) - 2F(s_3) + F(s_4)}\,, \tag{3}$$

where $F(s_1), F(s_2), F(s_3), F(s_4)$ denote the magnitude of DFT bins [1-3, 12, 13, 21].

Three new interpolated discrete Fourier transform (IpDFT) methods: the Yoshida-Magalas methods (the YM, the YM_C, and the Y_L [1-3]) and the original Yoshida method (Y) [12] use four DFT bins ($F(s_1), F(s_2), F(s_3), F(s_4)$) and a rectangular window [21]. The Y_L method differs from the Y method by the use of a fixed length of the signal $A(t)$ [1]. The YM method uses four optimal values of the DFT bins [1-3] whereas the YM_C differs from the YM method by using a complete number of oscillations [1-3]. The Yoshida method [12] and other IpDFT methods were recently reviewed in [13]. The systematic errors induced by spectral leakage and a picket fence effect were also discussed in [13]. Detailed mathematical description of IpDFT methods used in this work (i.e. the YM, the YM_C, and the Y_L) will be described elsewhere.

The performance of four IpDFT methods and the OMI method [1-8] for low damping level ($\delta = 5\times10^{-4}$, $f_0 = 1.12345$ Hz) is analyzed here for two sampling frequencies: $f_S = 1$ kHz and $f_S = 6$ kHz. It should be emphasized that the effect of the sampling frequency, f_S, was not investigated in the literature [4, 8].

The results of computing the logarithmic decrement δ for a set of 100 free decays are reported here as a function of the length of free decaying signals L [1-3, 5, 7] (in seconds and/or as a function of the number of oscillations L_{osc}) for two sampling frequencies, f_S. Each free decaying signal $A(t)$ was embedded in statistically different experimental noise defined by the same S/N ratio. The results of computations obtained from the OMI and IpDFT methods are offset independent [3, 7].

Figures 1, 2, 5, and 6 demonstrate that the OMI method (the results are illustrated by the 1st set of computed δ values vertically plotted from the left side) outperforms IpDFT methods: the YM, the YM_C, the Y_L, and the Y (the 2nd – 5th set of computed δ values vertically plotted) for all lengths of analyzed signals $A(t)$. It should be emphasized that the Yoshida method generates the highest dispersion in δ values and the highest relative error γ_δ (Figs. 2 and 6), the highest minimal $\gamma_{\delta\ min}$, and the maximal $\gamma_{\delta\ max}$ relative errors (Figs. 3, 7). The Y method returns the highest standard deviation too (Figs. 4, 8). It is convincingly demonstrated that the precision in computing the logarithmic decrement depends on the length of the signal while each method shows different performance. The Yoshida method returns strong dispersion for some well defined lengths of the signal $A(t)$ as a natural consequence of the relationship between the sampling and the resonant frequencies, and the fixed number of data points used in the algorithm [12] (e.g. Figs. 1, 5). That is why the sampling frequency f_S is a key factor which affects the performance of the Yoshida method, Y.

Computations of the δ for too short oscillations inevitably returns very strong dispersion (Figs. 1(a), 2(a) and 5(a), 6(a)). Is noteworthy that among four IpDFT methods the best performer is always the Yoshida-Magalas YM method [1-3]. Figures 1, 2 and 5, 6 indicate that the YM_C method should be used to compute the δ from short free decaying signals.

Figures 3 and 7 illustrate the performance of the IpDFT methods and the OMI method defined by the smallest relative error γ_δ, the smallest minimal $\gamma_{\delta\ min}$ and the maximal $\gamma_{\delta\ max}$ relative error in the estimation of the δ, that is, the smallest dispersion of experimental points in mechanical loss measurements.

An increase in the sampling frequency, from 1 kHz to 4 kHz, reduces the dispersion by around 50 %. Further increase up to 6 kHz yields the best estimation for the δ for all lengths of the free decaying signals and generates the smallest computing errors (compare Figs. 1 - 4 and Figs. 5 - 8).

3. Conclusions

The performance of different methods to compute the logarithmic decrement for low damping level ($\delta = 5\times10^{-4}$) can be listed in the following order: (1) the OMI, (2) the Yoshida-Magalas YM, (3) the YM_C, and the Yoshida (Y). The YM method outperforms other IpDFT methods [1-3, 12, 13] including the classic Yoshida method [12]. The YM method yields the smallest dispersion in experimental points of the logarithmic decrement δ for different lengths of free decaying oscillations and different sampling frequencies. The parametric OMI method is considered as the 'gold standard' in low-frequency high-resolution mechanical spectroscopy HRMS [1, 2]. It is emphasized that the sampling frequency is an important factor to obtain the lowest dispersion of experimental points, that is, the

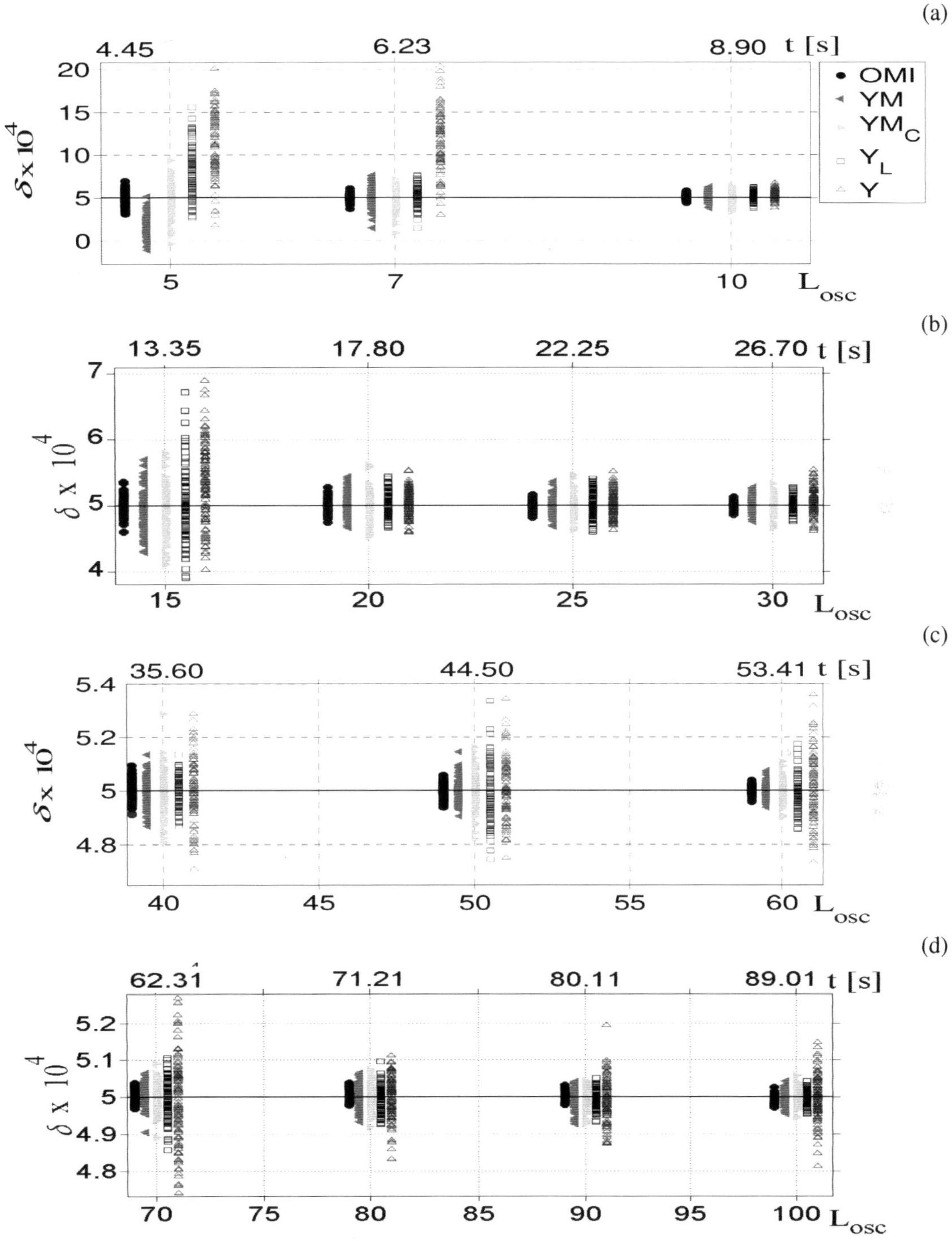

Figure 1. The effect of the sampling frequency f_S = 1 kHz on dispersion of 100 values of the logarithmic decrement δ computed according to OMI, YM, YM_C, Y_L, and Y methods as a function of the length of free decaying signals (i.e. the number of oscillations L_{osc} .) (a) L_{osc} = 5, 7, 10, (b) L_{osc} = 15, 20, 25, 30, (c) L_{osc} = 40, 50, 60, (d) L_{osc} = 70, 80, 90, and 100.

Computed values of the δ, displayed on vertical plots, correspond to a set of 100 different free decaying noisy oscillations (S/N = 32 dB) characterized by the same value of the δ = 0.0005 and the resonant frequency f_0 = 1.12345 Hz.

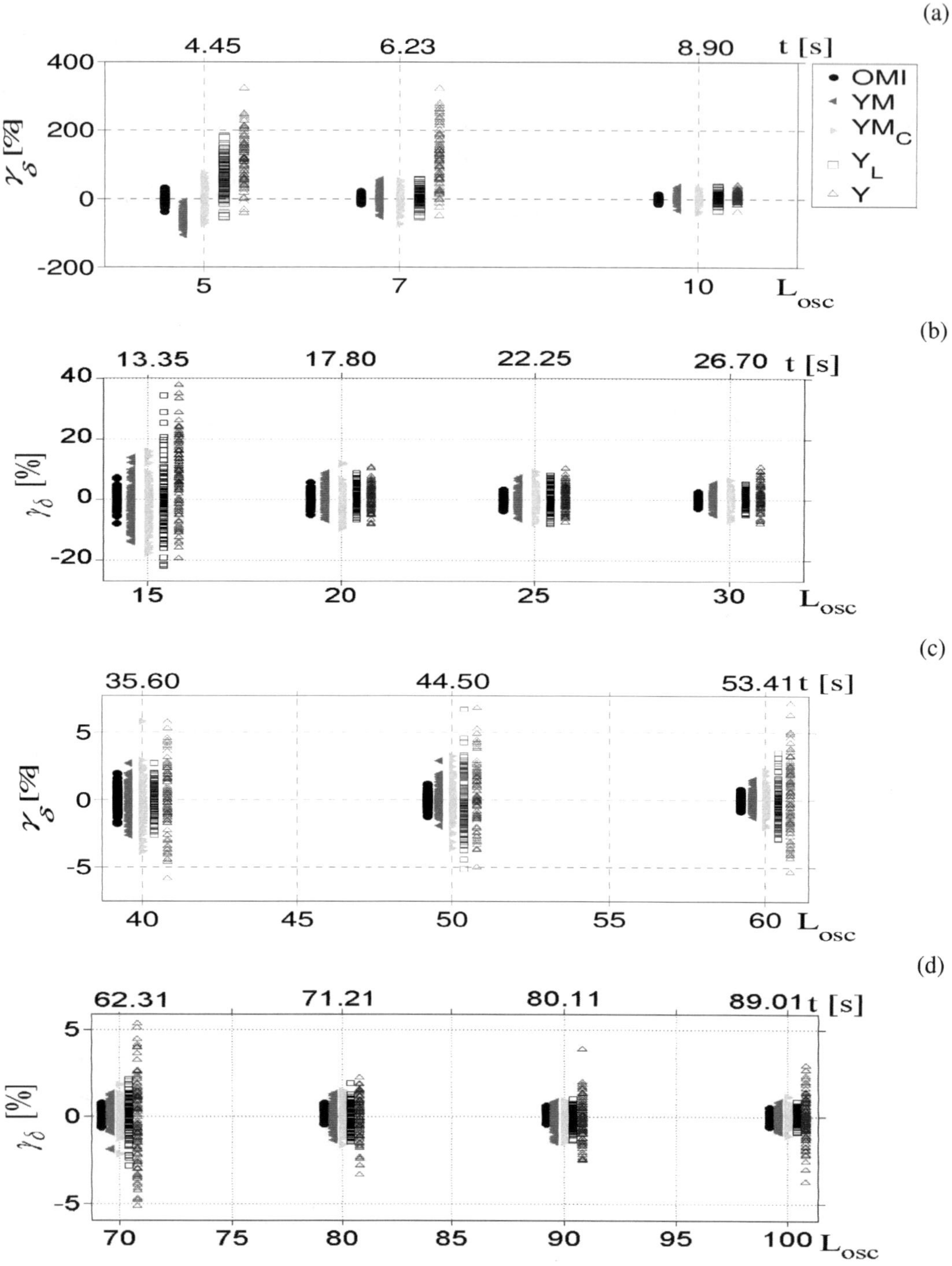

Figure 2. The effect of the sampling frequency f_S = 1 kHz on the relative errors γ_δ obtained for computations of the logarithmic decrement δ shown in Fig. 1 as a function of the length of free decaying signals (i.e. the number of oscillations L_{osc}.) (a) L_{osc} = 5, 7, 10, (b) L_{osc} = 15, 20, 25, 30, (c) L_{osc} = 40, 50, 60, (d) L_{osc} = 70, 80, 90, and 100.

(a)

(b)

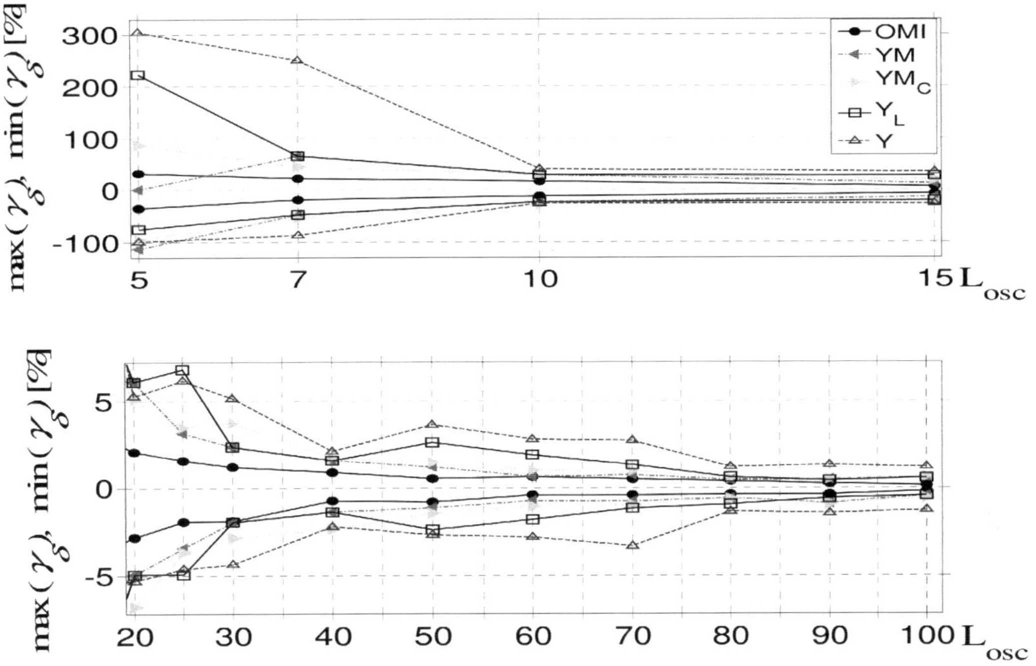

Figure 3. The effect of the sampling frequency f_S = 1 kHz on the minimal $\gamma_{\delta\,\min}$ and the maximal $\gamma_{\delta\ \max}$ relative errors obtained for computations of the logarithmic decrement δ shown in Fig. 1.
(a) L_{osc} = 5, 7, 10, 15, (b) L_{osc} = 20, 25, 30, 40, 50, 60, 70, 80, 90, and 100.

(a)

(b)

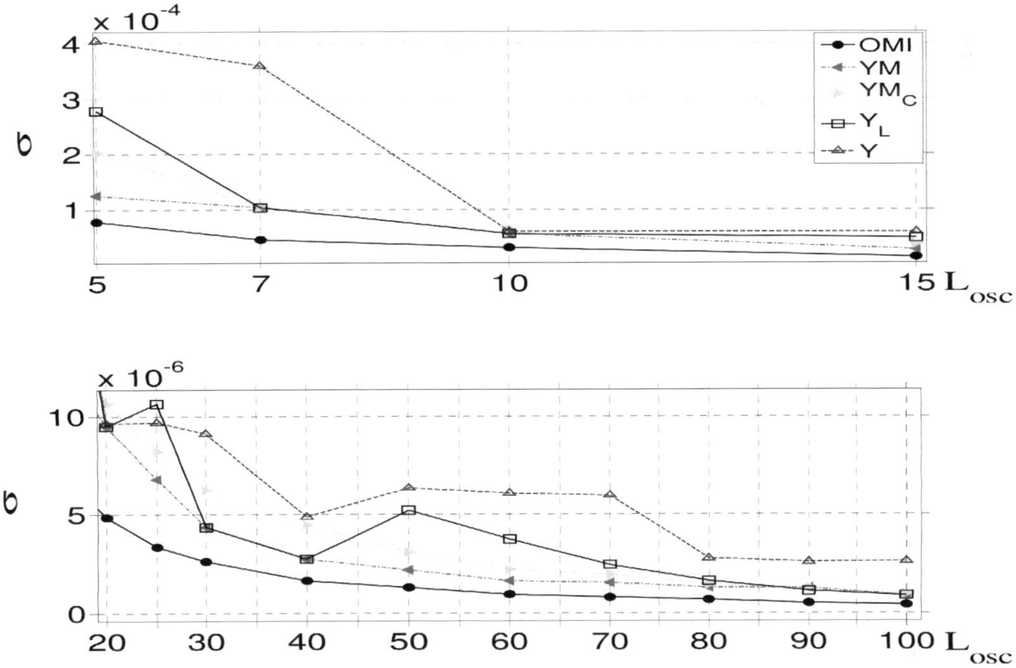

Figure 4. The effect of the sampling frequency f_S = 1 kHz on the standard deviation σ obtained for computations of the logarithmic decrement δ shown in Fig. 1. (a) L_{osc} = 5, 7, 10, 15, (b) L_{osc} = 20, 25, 30, 40, 50, 60, 70, 80, 90, and 100.

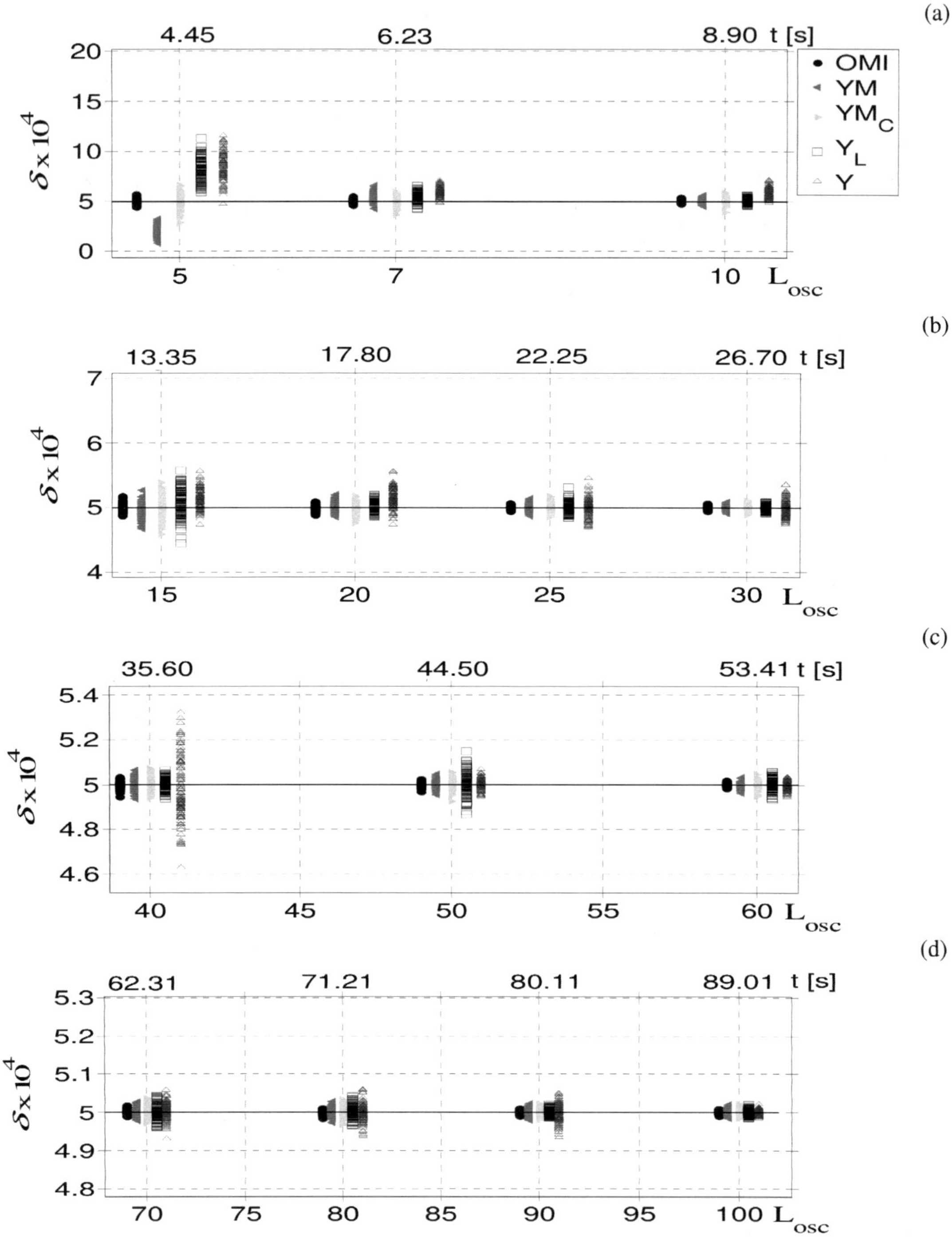

Figure 5. The effect of the sampling frequency $f_S = 6$ kHz on dispersion of 100 values of the logarithmic decrement δ computed according to OMI, YM, YM_C, Y_L, and Y methods as a function of the length of free decaying signals (i.e. the number of oscillations L_{osc} .) (a) L_{osc} = 5, 7, 10, (b) L_{osc} = 15, 20, 25, 30, (c) L_{osc} = 40, 50, 60, (d) L_{osc} = 70, 80, 90, and 100.

Computed values of the δ, displayed on vertical plots, correspond to a set of 100 different free decaying noisy oscillations (S/N = 32 dB) characterized by the same value of the δ = 0.0005 and the resonant frequency f_0 = 1.12345 Hz.

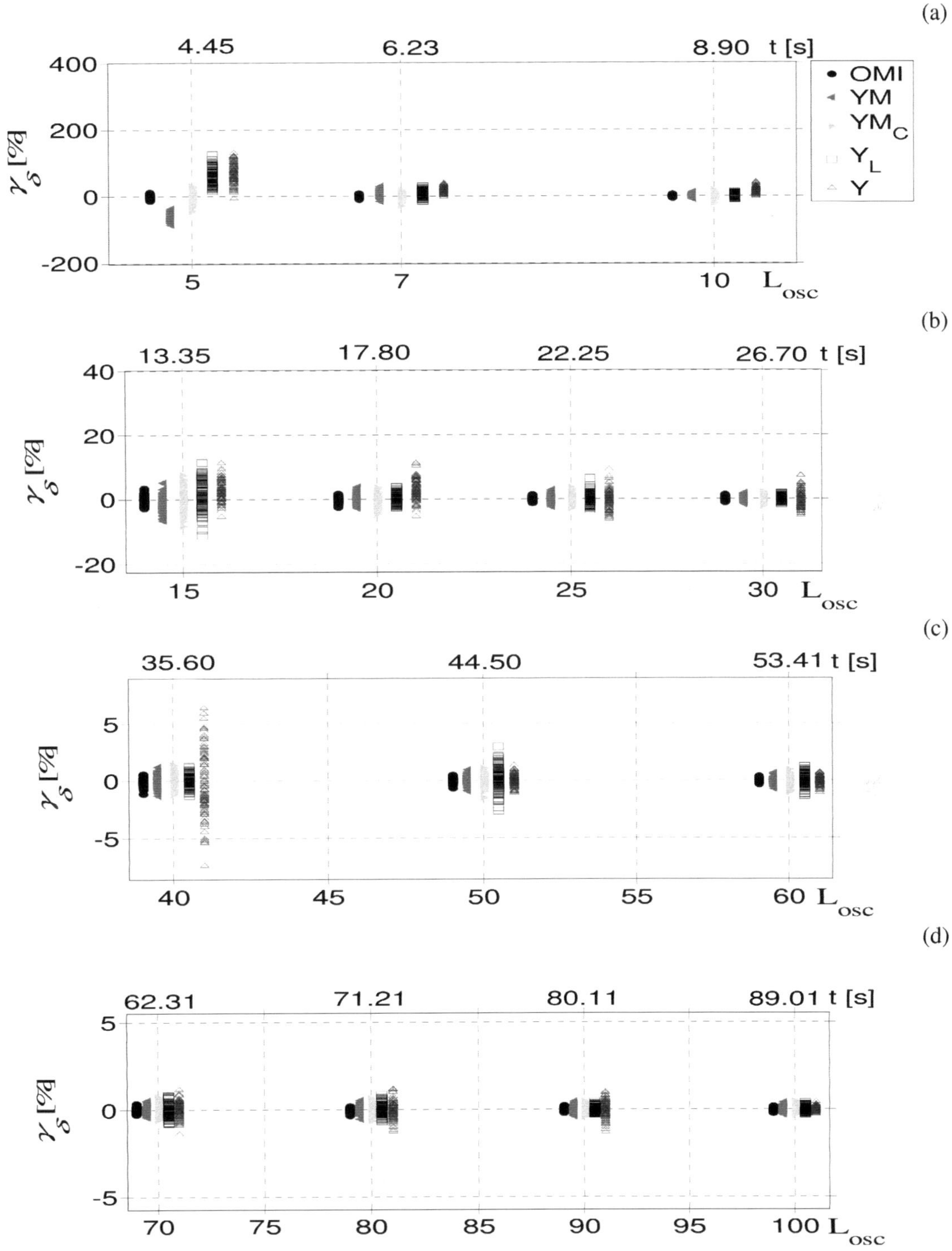

Figure 6. The effect of the sampling frequency f_S = 6 kHz on the relative errors γ_δ obtained for computations of the logarithmic decrement δ shown in Fig. 5 as a function of the length of free decaying signals (i.e. the number of oscillations L_{osc}.) (a) L_{osc} = 5, 7, 10, (b) L_{osc} = 15, 20, 25, 30, (c) L_{osc} = 40, 50, 60, (d) L_{osc} = 70, 80, 90, and 100.

IOP Conf. Series: Materials Science and Engineering **31** (2012) 012018 doi:10.1088/1757-899X/31/1/012018

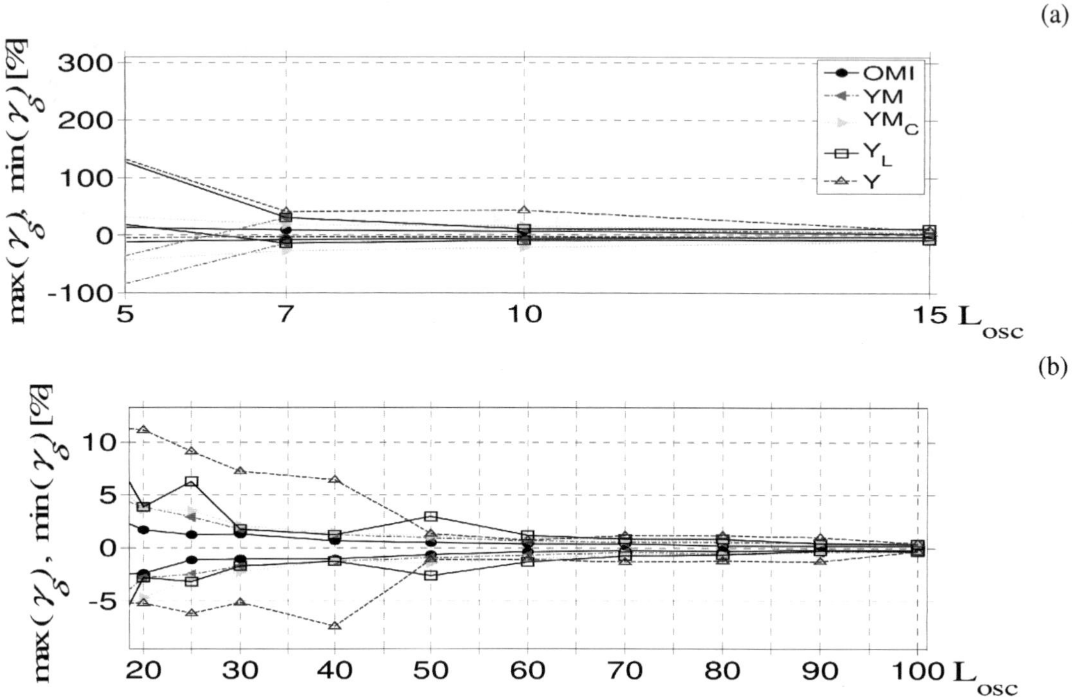

Figure 7. The effect of the sampling frequency f_S = 6 kHz on the minimal $\gamma_{\delta\,\min}$ and the maximal $\gamma_{\delta\ \max}$ relative errors obtained for computations of the logarithmic decrement δ shown in Fig. 5.
(a) L_{osc} = 5, 7, 10, 15, (b) L_{osc} = 20, 25, 30, 40, 50, 60, 70, 80, 90, and 100.

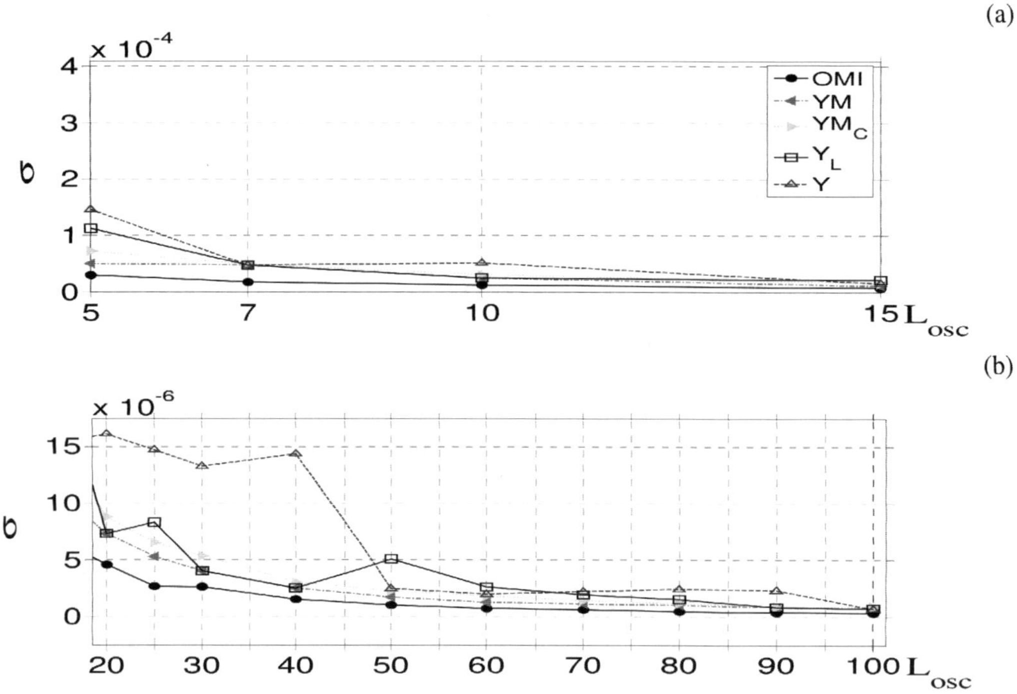

Figure 8. The effect of the sampling frequency f_S = 6 kHz on the standard deviation σ obtained for computations of the logarithmic decrement δ shown in Fig. 5.
(a) L_{osc} = 5, 7, 10, 15, (b) L_{osc} = 20, 25, 30, 40, 50, 60, 70, 80, 90, and 100.

lowest level of computing errors. It is concluded that the sampling frequency $f_S = 6$ kHz provides much better results as compared to usually used $f_S = 1$ kHz in low-frequency mechanical spectrometers operating around the resonant frequency $f_0 \approx 1$ Hz.

The OMI method and the Yoshida-Magalas YM method are recommended to compute the logarithmic decrement δ from exponentially damped time-invariant harmonic oscillations embedded in an experimental noise recorded in low-frequency mechanical spectrometers ($f_0 \approx 1$ Hz.) This conclusion is valid for low damping level ($\delta = 5\times10^{-4}$.)

Acknowledgements. This work was supported by Polish National Science Centre under grant No N N507 249040 and No N N507 446639.

References

[1] Magalas L B and Majewski M 2011 *Sol. St. Phen.* **184** 467-472
[2] Magalas L B and Majewski M 2011 *Sol. St. Phen.* **184** 473-478
[3] Majewski M 2011 *PhD Thesis*, AGH University of Science and Technology, Kraków, Poland
[4] Magalas L B 2006 *Sol. St. Phen.* **115** 7-14
[5] Magalas L B and Majewski M 2008 *Sol. St. Phen.* **137** 15-20
[6] Magalas L B and Malinowski T 2003 *Sol. St. Phen.* **89** 247-260
[7] Magalas L B and Majewski M 2009 *Mater. Sci. Eng.* A **521-522** 384-388
[8] Magalas L B and Stanisławczyk A 2006 *Key Eng. Materials* **319** 231-240
[9] Magalas L B and Piłat A 2006 *Sol. St. Phen.* 115 285-292
[10] Magalas L B 2003 *Sol. St. Phen.* **89** 1-22
[11] Etienne S, Elkoun S, David L and Magalas L B 2003 *Sol. St. Phen.* **89** 31-66
[12] Yoshida I, Sugai T, Tani S, Motegi M, Minamida K and Hayakawa H 1981 *J. Phys. E: Sci. Instrum.* **14** 1201-1206
[13] Duda K, Magalas L B, Majewski M and Zieliński T P 2011 *IEEE Transactions on Instrumentation and Measurement* **60** 3608-3618
[14] Magalas L B, Dufresne J F and Moser P 1981 *J. de Phys.* **42** (C-5) 127-132
[15] Magalas L B, Moser P and Ritchie I G 1983 *J. de Phys.* **44** (C-9) 645-649
[16] Magalas L B and Gorczyca S 1985 *J. de Phys.* **46** (C-10) 253-256
[17] Rubianes J, Magalas L B and Fantozzi G 1987 *J. de Phys.* **48** (C-8) 185-190
[18] Ngai K L, Wang Y N and Magalas L B 1994, *J. Alloy Compd.* **211/212** 327-332
[19] Magalas L B 1996 *J. de Phys. IV* **6** (C8) 163-172
[20] Magalas L B 2003 *Acta Metallurgica Sinica* **39** 1145-1152
[21] Brigham E O 1988 *The Fast Fourier Transform FFT and its Applications*, Prentice Hall Signal Processing Series

IOP Conf. Series: Materials Science and Engineering **31** (2012) 012019 doi:10.1088/1757-899X/31/1/012019

Advances in Computational High-Resolution Mechanical Spectroscopy HRMS. Part II - Resonant Frequency – Young's Modulus

M Majewski and L B Magalas

AGH University of Science and Technology, Faculty of Metals Engineering and Industrial Computer Science, al. Mickiewicza 30, 30-059 Kraków, Poland

E-mail: magalas@agh.edu.pl

Abstract. In this paper, we compare the values of the resonant frequency f_0 of free decaying oscillations computed according to the parametric OMI method (*Optimization in Multiple Intervals*) and nonparametric DFT-based (discrete Fourier transform) methods as a function of the sampling frequency. The analysis is carried out for free decaying signals embedded in an experimental noise recorded for metallic samples in a low-frequency resonant mechanical spectrometer. The Yoshida method (Y), the Agrež method (A), and new interpolated discrete Fourier transform (IpDFT) methods, that is, the Yoshida-Magalas (YM) and (YM_C) methods developed by the authors are carefully compared for the resonant frequency f_0 = 1.12345 Hz and the logarithmic decrement, δ = 0.0005. Precise estimation of the resonant frequency (Youngs' modulus ~ f_0^2) for real experimental conditions, i.e., for exponentially damped harmonic signals embedded in an experimental noise, is a complex task. In this work, various computing methods are analyzed as a function of the sampling frequency used to digitize free decaying oscillations. The importance of computing techniques to obtain reliable and precise values of the resonant frequency (i.e. Young's modulus) in materials science is emphasized.

1. Introduction

High-resolution mechanical spectroscopy HRMS [1, 2, 12] provides excellent estimation of Young's modulus E, which is proportional to the square of the resonant frequency of harmonically oscillating sample, E ~ f_0^2 [3-6]. This paper focuses on the performance and the accuracy of discrete Fourier transform-based (DFT-based) methods [19-22] to obtain the resonant frequency f_0 from exponentially damped harmonic oscillations (i.e., free decaying oscillations) recorded in a low-frequency mechanical spectrometer [7]. The parametric OMI method (*Optimization in Multiple Intervals*) [8-12] and interpolated discrete Fourier transform (IpDFT) techniques [1, 2, 12-16] are used to analyze exponentially damped harmonic oscillations characterized by the following parameters: the length of the free decaying signal L, the sampling frequency f_S, the signal-to-noise ratio S/N, phase, the resonant frequency f_0, and the damping level (i.e. the logarithmic decrement, δ). The discrete Hilbert transform-based methods are not discussed in this work [6, 17, 19].

The OMI method [8] and IpDFT methods [12-15] compute jointly the logarithmic decrement and the resonant frequency. Therefore the problem of the computations of the δ and the f_0 cannot be discussed separately. That is why, the results concerning the δ and the f_0 are reported in this volume in two Parts.

The effect of the Zero-Point Drift (ZPD) [11, 12, 17, 18] on computations of the resonant frequency is not discussed in this work. The OMI method and the IpDFT methods are not affected by the presence of constant offsets, unlike the classical algorithms reported in the literature [3, 4, 8]. A new road to high-resolution mechanical spectroscopy takes into consideration the presence of the ZPD for all damping levels and different lengths of the signal [7, 12, 18]; these results will be reported elsewhere.

2. Resonant Frequency

The exponentially damped time-invariant harmonic oscillations embedded in an experimental noise $\varepsilon_w(t)$ can be described using the digitized data $A_i(t)$ and t_i from free decaying signal $A(t)$ [1, 2]:

$$A(t) = A_0 e^{-\delta f_0 t} \cos(2\pi f_0 t + \varphi) + \varepsilon_w(t) + dc\,, \tag{1}$$

where A_0 is the maximal strain amplitude of a sample mounted in a mechanical spectrometer, t is a continuous time in seconds, $-\pi < \varphi \le \pi$ is the phase of the signal $A(t)$ in radians, f_0 is the resonant frequency, and dc is an offset. The noise $\varepsilon_w(t)$ corresponds here to the signal-to-noise ratio S/N= 32 dB [1, 2, 8, 10, 12].

It can be shown that the resonant frequency f_0 can be directly computed from the DFT spectra:

$$f_0 = \frac{Re\left(s_1 - \dfrac{3}{R-1}\right)}{L}. \tag{2}$$

The R parameter is defined by Yoshida [13]:

$$R = \frac{F(s_1) - 2F(s_2) + F(s_3)}{F(s_2) - 2F(s_3) + F(s_4)}, \tag{3}$$

where $F(s_1), F(s_2), F(s_3), F(s_4)$ state for the magnitude of DFT bins [1, 2, 12, 13, 15].

The Yoshida-Magalas methods (YM) and (YM_C) [1, 2, 12] and the original Yoshida method (Y) [13] use four DFT bins ($F(s_1), F(s_2), F(s_3), F(s_4)$) and a rectangular window [19, 22]. The YM method uses four optimal values of the DFT bins whereas the YM_C differs from the YM method by using a complete number of oscillations. The Yoshida method is described in [13]. The Agrež method (A) uses three DFT bins [14] and the Hann window [20, 22]. The wavelet transform gives too poor frequency resolution [16, 23-26], and is not discussed in this work.

3. Results and Discussion

The results of computing the resonant frequency f_0 according to Eqs. (2, 3) by the YM, the YM_C [1, 2, 12], and the Y methods [13] are carefully compared here with the results obtained according to the A method [14] and the parametric OMI method [8-12]. The results shown in Figs. 1 - 4 confirm that the OMI method can be considered as the 'gold standard' [1, 2, 7-12, 15].

Figure 1 shows dispersion of computed f_0 values obtained for 100 free decaying oscillations embedded in an experimental noise as a function of the length of signals, L, (in seconds and/or as a function of the number of oscillations L_{osc}) [1, 2, 10-12] for the sampling frequency $f_S = 1$ kHz. The dispersion of computed f_0 values obtained for the sampling frequency $f_S = 6$ kHz is illustrated in Fig. 3. The results obtained for 100 different free decaying oscillations ($\delta = 5\times10^{-4}$, $f_0 = 1.12345$ Hz, S/N = 32 dB) computed according to the following methods: OMI, YM, YM_C, A, and Y are vertically plotted in Figs. 1, 2 and Figs. 3, 4 for $f_S = 1$ kHz and $f_S = 6$ kHz, respectively.

The OMI unequivocally outperforms IpDFT methods for all lengths of free decaying signals. The Yoshida method [13] usually generates the highest dispersion in f_0 points (see Figs. 1-4.) It is noteworthy that the dispersion in f_0 values increases with decrease in the length of free decaying signals. This observation is valid for all IpDFT methods and all damping levels. The Yoshida-Magalas YM method returns the smallest dispersion in experimental points among all tested IpDFT methods. The point we should like to emphasize is that computed f_0 values are biased for signals that are too short (Figs. 1(a), 2(a), 2(c), 3(a), 4(a), 4(c).)

Figures 2 and 4 show variation of the minimal $\gamma_{f_o\ \min}$ and the maximal $\gamma_{f_o\ \max}$ relative errors of the resonant frequency f_0 for two sampling frequencies: $f_S = 1$ kHz and $f_S = 6$ kHz, respectively. The computed values of the f_0 are confined within two lines: $\gamma_{f_o\ \min}$ and $\gamma_{f_o\ \max}$ corresponding to the specific method. The method which returns the best f_0 values returns simultaneously the best results for the logarithmic decrement (see Part I in this volume.)

An increase in the sampling frequency f_S from 1 kHz to 4 kHz reduces relative errors by 5-15 %. Further increase in the sampling frequency returns slightly better results only.

It is concluded that the performance of tested methods is as follows: (1) OMI – it is worthwhile to reiterate the fact that the OMI is considered as the 'gold standard', (2) the Yoshida-Magalas YM, (3) the Yoshida-Magalas YM_C, (4) the Agrež A, and (5) the Yoshida Y. The results shown in Figs. 1-4 suggest that for short and very short signals only the OMI and the YM_C can be recommended. The Agrež method, A, provides good estimation of the resonant frequency for low damping level only. For medium and higher damping levels it returns low quality unacceptable results. That is why the A method can only be used in the computations of the f_0 for relatively narrow span of low damping level. It is important to emphasize that in all investigated cases the Yoshida-Magalas YM method is slightly better as compared to the Agrež method. The potential use of the Agrež method [14] in mechanical spectroscopy and other spectroscopic techniques will be limited since it yields wrong results for the logarithmic decrement, δ.

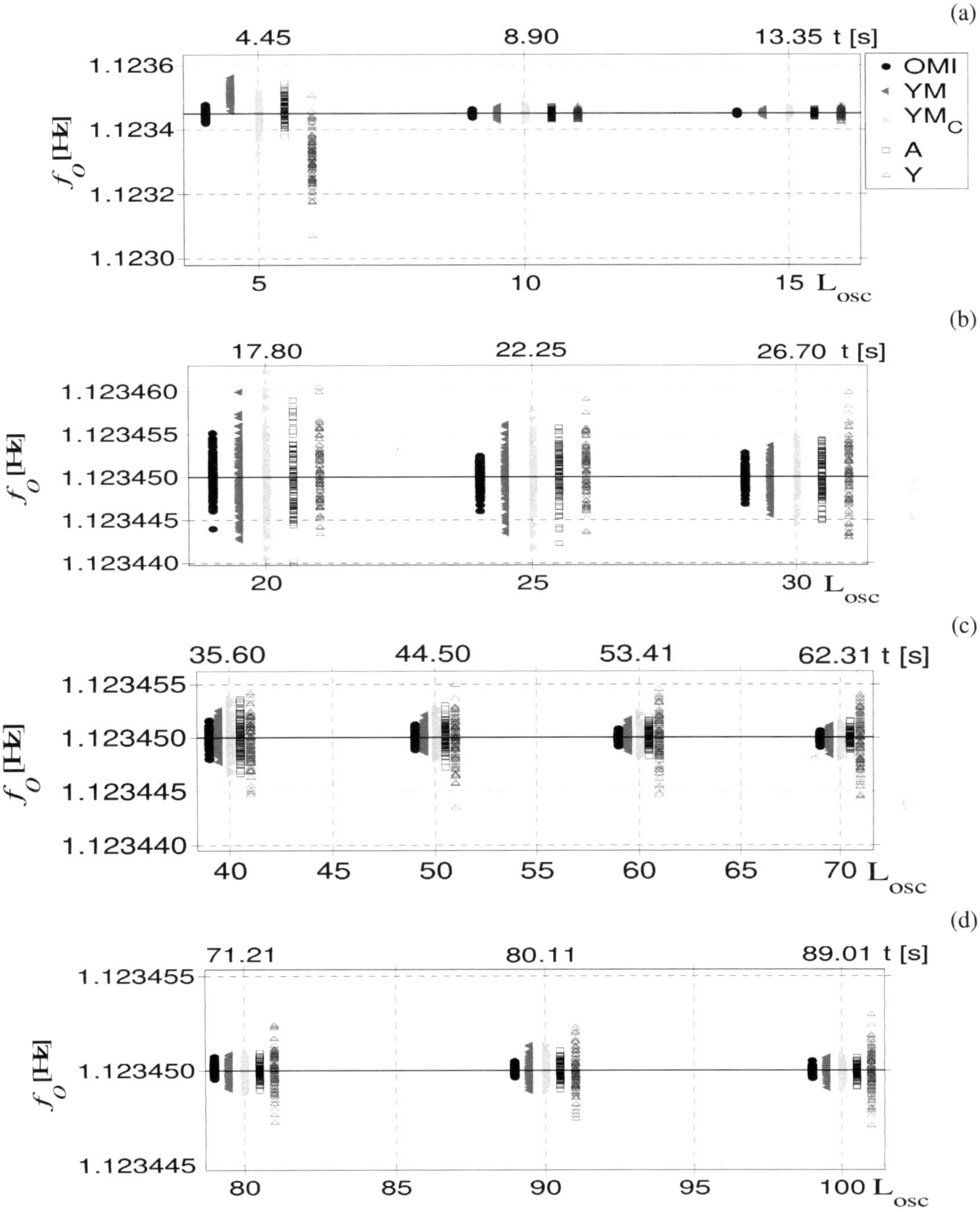

Figure 1. The effect of the sampling frequency f_S = 1 kHz on dispersion of 100 values of the resonant frequency f_0 computed according to OMI, YM, YMc, A, and Y methods as a function of the length of free decaying signals (i.e. the number of oscillations L_{osc}.) (a) L_{osc} = 5, 10, 15, (b) L_{osc} = 20, 25, 30, (c) L_{osc} = 40, 50, 60, 70, (d) L_{osc} = 80, 90, and 100.
Computed values of the f_0, displayed on vertical plots, correspond to a set of 100 different free decaying noisy oscillations (S/N = 32 dB) characterized by the same value of the δ = 0.0005 and the resonant frequency f_0 = 1.12345 Hz.

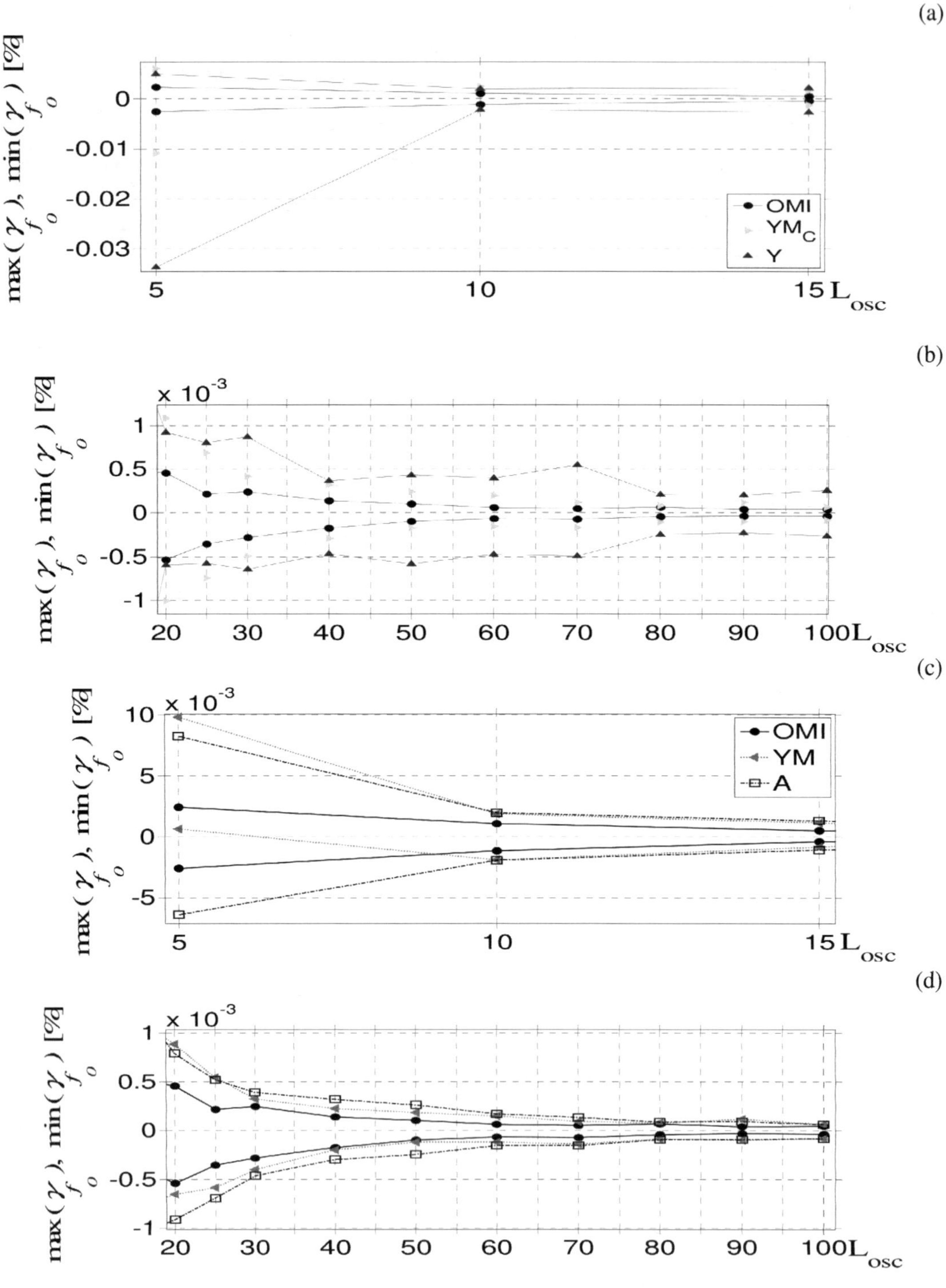

Figure 2. The effect of the sampling frequency f_S = 1 kHz on the minimal $\gamma_{f_0 \min}$ and the maximal $\gamma_{f_0 \max}$ relative errors obtained for computations of the resonant frequency f_0 shown in Fig. 1.
(a) , (c) L_{osc} = 5, 10, 15; (b), (d) L_{osc} = 20, 25, 30, 40, 50, 60, 70, 80, 90, and 100.

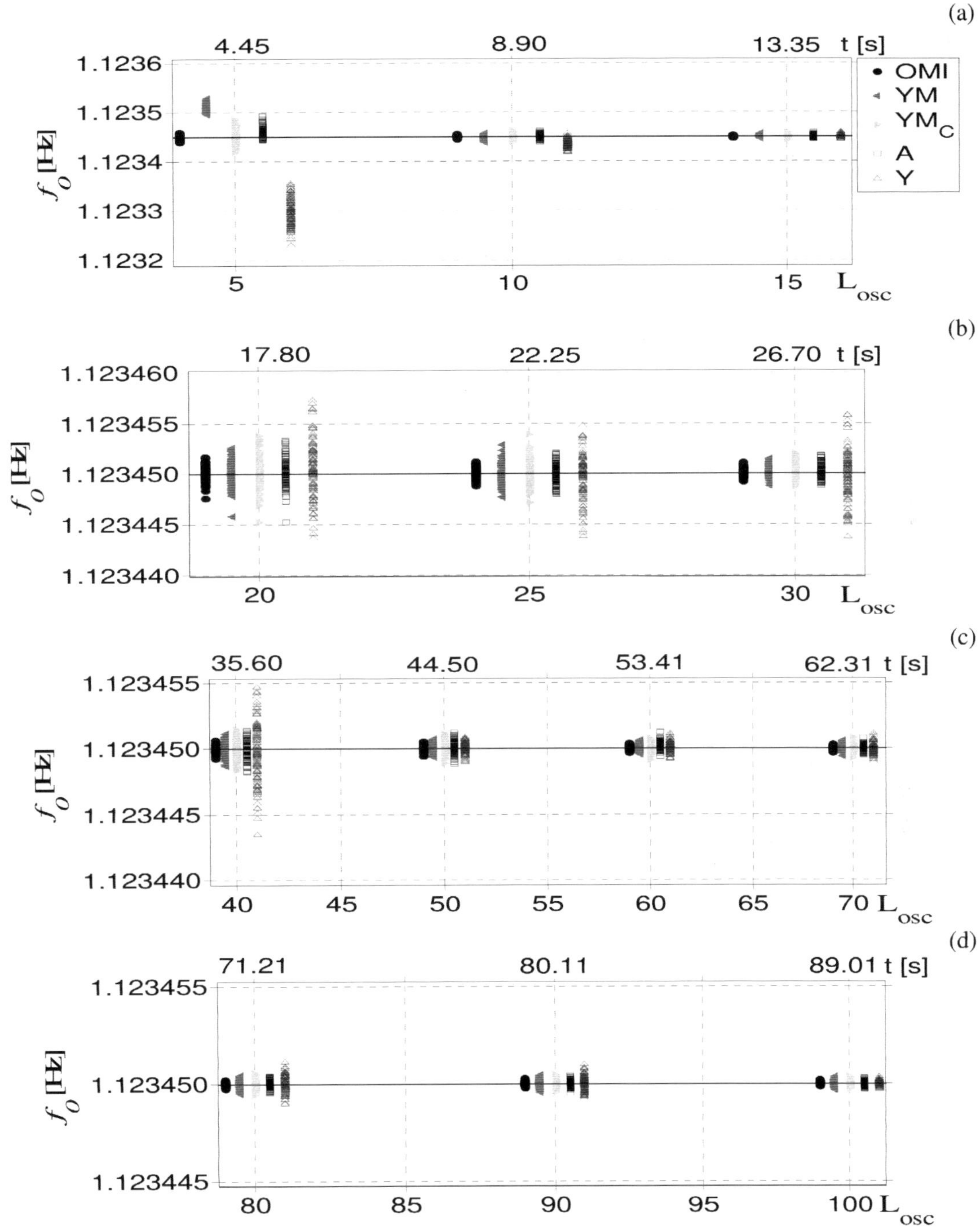

Figure 3. The effect of the sampling frequency f_S = 6 kHz on dispersion of 100 values of the resonant frequency f_0 computed according to OMI, YM, YMC, A, and Y methods as a function of the length of free decaying signals (i.e. the number of oscillations L_{osc} .) (a) L_{osc} = 5, 10, 15, (b) L_{osc} = 20, 25, 30, (c) L_{osc} = 40, 50, 60, 70, (d) L_{osc} = 80, 90, and 100.

Computed values of the f_0, displayed on vertical plots, correspond to a set of 100 different free decaying noisy oscillations (S/N = 32 dB) characterized by the same value of the δ = 0.0005 and the resonant frequency f_0 = 1.12345 Hz.

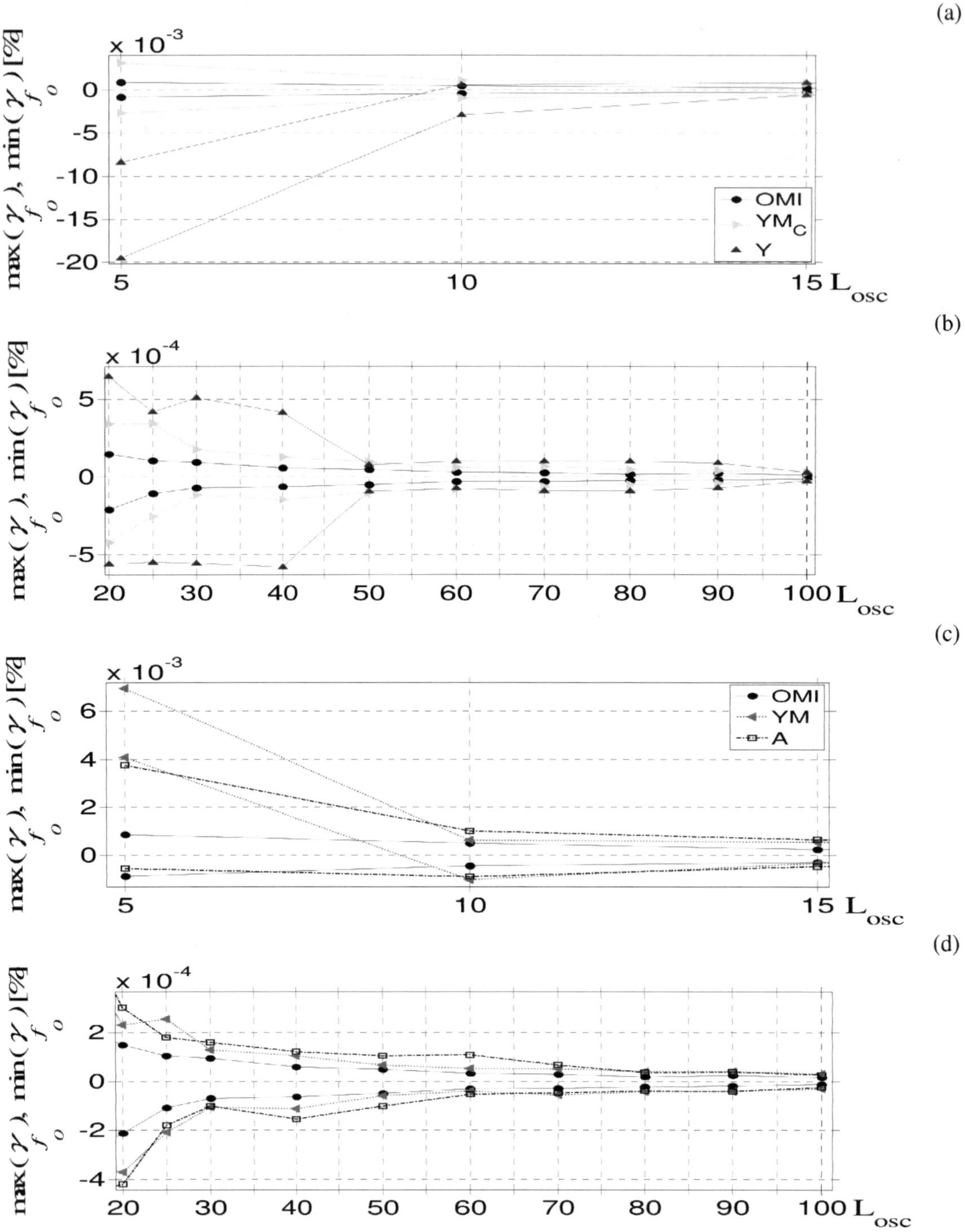

Figure 4. The effect of the sampling frequency f_S = 6 kHz on the minimal $\gamma_{f_0 \min}$ and the maximal $\gamma_{f_0 \max}$ relative errors obtained for computations of the resonant frequency f_0 shown in Fig. 3.
(a) , (c) L_{osc} = 5, 10, 15; (b), (d) L_{osc} = 20, 25, 30, 40, 50, 60, 70, 80, 90, and 100.

4. Conclusions

The results are primarily reported here to highlight new methods to compute the resonant frequency, f_0, and the logarithmic decrement, δ, in resonant high-resolution mechanical spectroscopy, HRMS. The performance of different methods and algorithms to compute the resonant frequency for low damping level (e.g. $\delta = 5\times10^{-4}$) is listed in the following order: (1) the OMI, (2) the Yoshida-Magalas YM, (3) the Yoshida-Magalas YM_C, (4) the Agrež A, and finally (5) the Yoshida Y. The Yoshida-Magalas YM method [1, 2] outperforms other IpDFT methods. The YM method yields the smallest dispersion in experimental points of the resonant frequency for different lengths of free decaying oscillations and different sampling frequencies. The effect of the sampling frequency on precision in the computations of the resonant frequency and the logarithmic decrement has received scant attention to date [8] and deserves more.

The parametric OMI method is considered as the 'gold standard' in low-frequency high-resolution mechanical spectroscopy, HRMS. The point we should like to emphasize is that the sampling frequency is a key factor to reduce dispersion of experimental points computed according to the OMI and IpDFT methods. It is demonstrated that the sampling frequency $f_S = 6$ kHz readily yields better results as compared to usually used $f_S = 1$ kHz in low-frequency mechanical spectrometers operating around the resonant frequency $f_0 \approx 1$ Hz.

To conclude, the OMI method (*Optimization in Multiple Intervals*) and the Yoshida-Magalas YM method are recommended to compute the resonant frequency from the exponentially damped time-invariant harmonic oscillations embedded in an experimental noise. It is not difficult to show, by means of the experimental results reported in [1, 2] and the results described in this volume that the OMI and the YM methods pave the way toward high-resolution mechanical spectroscopy, HRMS.

Acknowledgements. This work was supported by Polish National Science Centre under grant No N N507 249040 and No N N507 446639.

References

[1] Magalas L B and Majewski M 2011 *Sol. St. Phen.* **184** 467-472
[2] Magalas L B and Majewski M 2011 *Sol. St. Phen.* **184** 473-478
[3] Nowick A S and Berry B S 1972 *Anelastic Relaxation in Crystalline Solids*, Academic Press, New York and London
[4] Magalas L B 2003 *Sol. St. Phen.* **89** 1-22
[5] Etienne S, Elkoun S, David L and Magalas L B 2003 *Sol. St. Phen.* **89** 31-66
[6] Magalas L B and Malinowski T 2003 *Sol. St. Phen.* **89** 247-260
[7] Magalas L B and Malinowski T 2003 *Sol. St. Phen.* **89** 349-354
[8] Magalas L B 2006 *Sol. St. Phen.* **115** 7-14
[9] Magalas L B and Stanisławczyk A 2006 *Key Eng. Materials* **319** 231-240
[10] Magalas L B and Majewski M 2008 *Sol. St. Phen.* **137** 15-20
[11] Magalas L B and Majewski M 2009 *Mater. Sci. Eng.* A **521-522** 384-388
[12] Majewski M 2011 *Phd Thesis*, AGH University of Science and Technology, Kraków, Poland
[13] Yoshida I, Sugai T, Tani S, Motegi M, Minamida K and Hayakawa H 1981 *J. Phys. E: Sci. Instrum.* **14** 1201-1206
[14] Agrež D 2009 *IEEE Instrumentation and Measurement Technology Conference* **1-3** 1295-1300
[15] Duda K, Magalas L B, Majewski M and Zieliński T P 2011 *IEEE Transactions on*

Instrumentation and Measurement **60** 3608-3618
[16] Magalas L B 2000 *J. Alloy Compd.* **310** 269-275
[17] Magalas L B 1996 *J. de Phys. IV* **6** (C8) 163-172
[18] Magalas L B and Piłat A 2006 *Sol. St. Phen.* **115** 285-292
[19] Poularikas A D ed. 1996 *The Transforms and Applications,* CRC Press Inc.
[20] Brigham E O 1988 *The Fast Fourier Transform FFT and its Applications*, Prentice Hall Signal Processing Series
[21] Ramirez R W 1985 *The FFT Fundamentals and Concepts*, Prentice-Hall International Inc.
[22] Oppenheim A V, Schafer R W and Buck J R 1999 *Discrete–Time Signal Processing*, Prentice–Hall
[23] Young Randy K 1993 *Wavelet Theory and its Applications,* Kluwer Academic Publishers
[24] Chan Y T 1995 *Wavelet Basics*, Kluwer Academic Publishers
[25] Newland D E 1993 *Random Vibrations, Spectral and Wavelet Analysis,* Longman Scientific & Technical
[26] Qian S and Chen D 1996 *Joint Time-Frequency Analysis*, The MathWorks, Inc.

IOP Conf. Series: Materials Science and Engineering **31** (2012) 012020 doi:10.1088/1757-899X/31/1/012020

Simultaneous measurements of thermal conductivity and electrical conductivity of micro-machined Silicon films

H Hagino[1], Y Kawahara[1], A Goto[1], and K Miyazaki[1]
Department of Mechanical Engineering, Kyushu Institute of Technology, Sensui-cho, Tobata-ku, Kitakyushu, JAPAN

E-mail: j344138h@tobata.isc.kyutech.ac.jp

Abstract. The in-plane effective thermal conductivity of free-standing Si thin films with periodic micropores was measured at -100 to 0 °C. The Si thin films with micropores were prepared from silicon-on-insulator (SOI) wafers by standard microfabrication processes. The dimensions of the free-standing Si thin films were 200μm×150μm×2 μm, with staggered 4 μm pores having an average pitch of 4 μm. The Si thin film serves both as a heater and thermometer. The average temperature rise of the thin film is a function of its in-plane thermal conductivity. The effective thermal conductivity was calculated using a simple one-dimensional heat conduction model. The measured thermal conductivity was much lower than that expected based on classical model evaluations. A significant phonon size effect was observed even in the microsized structures, and the mean free path for phonons is very long even at the room temperature.

1. Introduction

Thermal energy is transported in solids by mainly two carriers; electrons in metals, and phonons in insulators and semiconductors. Two fundamental length scales of wavelength and the mean free path (MFP) for carriers become significant in the thermal management of microstructured solids, such as an integrated circuit. The number of elements in integrated circuits is increasing daily in accordance with Moore's Law [1]. Currently there are more than 10 billion elements in large scale integrated circuits. Modern manufacturing techniques allow us to fabricate microstructure dimensions with sizes in the 1-100 nm range. The phonon MFP is longer than that expected based on the simple phonon gas model, and the MFP is more than 100 nm [2]. Artificially reduced thermal conductivity has been reported using nanostructured materials [3]. One of the applications of thermal conductivity reduction is an enhancement of thermoelectric performance [4,5]. Thermoelectric materials convert low grade waste heat into electricity, which makes it a key technology that contributes to sustainability through the scavenging of waste heat. The efficiency of thermoelectric materials is calculated as the thermoelectric figure of merit $ZT(=\sigma S^2T/\lambda)$, which is defined by the electrical conductivity σ, the Seebeck coefficient S, the temperature T, and the thermal conductivity λ. Thermoelectric conversion has been too inefficient to be cost effective in most applications. Therefore, many attempts have been made to enhance the figure of merit using nanostructured materials over the past 20 years [4,5]. It has been increasingly recognized that thermal conductivity in nanostructured materials is not an intrinsic parameter as it is for bulk materials [6, 7]. In particular, Si nanowires have shown that significant

[1] Department of Mechanical Engineering, Kyushu Institute of Technology, Sensui-cho, Tobata-ku, Kitakyushu, JAPAN

IOP Conf. Series: Materials Science and Engineering **31** (2012) 012020 doi:10.1088/1757-899X/31/1/012020

reduction of the thermal conductivity is induced by strong phonon scattering without any major decrease of electrical conductivity [8]. The operation of Si nanowires for thermoelectric applications is based on the large difference in the MFP lengths of electrons and phonons at room temperature. In this study, we propose design guidelines for a reduction in the thermal conductivity by evaluating the MFP of phonons. Free-standing Si thin films with periodic microholes were prepared, where the Si thin film serves as both a heater and a thermometer. The measured temperature rise of the self-heated thin film is a function of its in-plane effective thermal conductivity. We discuss the measured thermal conductivity of microstructured Si with respect to the MFP of phonons.

2. Measurement model

Figure 1 shows a schematic of the measurement system, the physical model, and the coordinate system for the following theoretical analysis. A Si thin film of length L is supplied with a constant direct current to generate a uniform heat flux; it is used as a heater for self-heating. The temperature distribution is assumed to be the state shown in figure 1. The temperature distribution is defined by the following equation from the one-dimensional heat conduction equation in steady state.

$$\frac{\partial^2 T}{\partial x^2} = -\frac{q_v}{c\rho\alpha} = -\frac{q_v}{k} \tag{1}$$

where q_v is the heat generation per volume unit, c is the specific heat, ρ is the degree of density, α is the thermal diffusion, and κ is the thermal conductivity. T_0 is defined as the temperature of the terminus.

$$T = -\frac{q_v}{2\lambda}(x^2 - Lx) + T_0 \tag{2}$$

T_{AVE} is the average temperature and is calculated by integrating the interval L.

$$T_{AVE} = \frac{1}{L}\int_0^L T dx = T_0 + \frac{qL^2}{12\kappa} \tag{3}$$

The thermal conductivity equation is obtained as follows [9]:

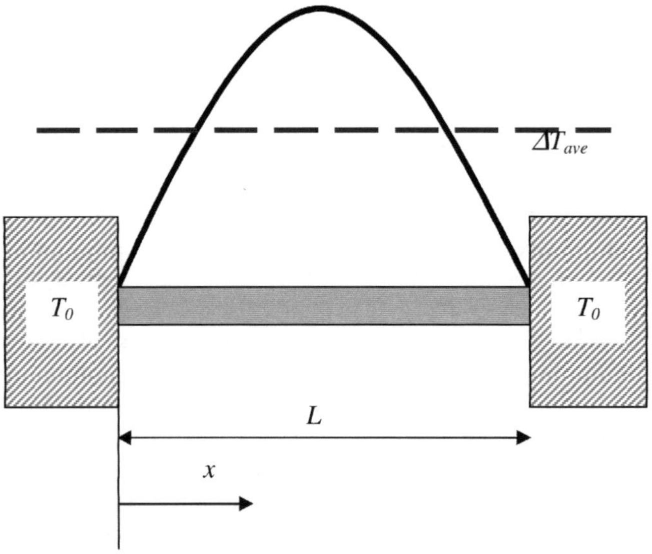

Figure 1. Physical model and temperature distribution of a Si microbridge during self-heating.

$$\kappa = \frac{qL^2}{12\left(T_{AVE} - T_0\right)} \tag{4}$$

The electrical conductivity is calculated as follows:

$$\sigma = \frac{L}{RA} \tag{5}$$

Here, σ is the electrical conductivity and R is the electrical resistance and L/A is the cross sectional area of the Si thin film.

3. Experimental

3.1. Sample preparation

Free-standing Si thin films with and without straggled microholes were fabricated using microfabrication processes. The fabrication processes involved repeated deposition, photolithography and etching and developing of single Si and oxide-based silicon-on-insulator (SOI) wafers. The process steps for the fabrication of the thin film single {110} wafers are shown in figure 2. The natural oxide film on the surface was removed using dilute HF acid and first etching was conducted in KOH (34 wt%) to obtain a surface Si thickness of 2 μm (figure 2(a)). The free-standing Si thin film pattern was printed using a lithographic exposure tool on a mask to a thin layer of photosensitive material covering the surface of the single Si layer (figure 2(b)). The surface of the Si thin film was removed by reactive ion etching (RIE) (figure 2(c)). The microholes were fabricated in the RIE process. Ion implantation was performed to fabricate electrodes. The SiO_2 was grown by chemical vapor deposition (CVD) and the patterned surface SiO_2 was removed by RIE for ion implantation (figure 2(d)). Phosphorus was implanted into the Si surface using a medium-current ion implanter with 1.0×10^{15} ions at 30 kV. The SiO_2 mask for ion implantation was removed and the sample was annealed at 950 °C for 30 min in nitrogen atmosphere for ion diffusion. The backside Si was removed by wet chemical etching using KOH (34 wt%) followed by growth of a SiN_x thin film by CVD for protection of the free-standing Si thin film (figure 2(e)). Both SiO_2 and SiN_x films around the free-standing Si were removed using buffered fluoride (BF; HF =1:6). Finally, an Al thin film was deposited on the electrodes of the free-standing Si thin film and annealed at 450 °C for 90 min in an Ar atmosphere. The dimensions of the free-standing Si thin films 200×150×2 μm^3 with staggered 4 μm pores and an average pitch of 4 μm, as shown in figure 3.

3.2. Temperature measurement.

The average temperature of the free-standing Si thin film T_{AVE} was determined based on the temperature dependence of electrical resistance of the Si thin film. The electrical resistance was measured from -100 to 60 °C using the 4 probe method for the calibration. Figure 4 shows the temperature dependence of the electrical resistance, where R_0 is the electrical resistance of the free-standing Si thin film and R_s is that of the thin film with periodic staggered pores. The electrical resistance of both samples increased with increasing temperature and was reduced near room temperature. The results are consistent with the well known temperature dependency of the electrical resistance of a semiconductor according to the carrier concentration and mobility.

3.3. Thermal conductivity measurement

Figure 5 shows a schematic diagram of the experimental apparatus for thermal conductivity measurement in a vacuum chamber. The temperature at the edge of the film T_0 was measured using a thermocouple fixed near the edge of the thin film. DC 2 V was applied to the free-standing Si thin film and the electric current and voltage were measured in the range from -100 to 0 °C during joule heating of the Si thin film. The thermal conductivity was calculated from T_{ave} measured from the electrical resistance

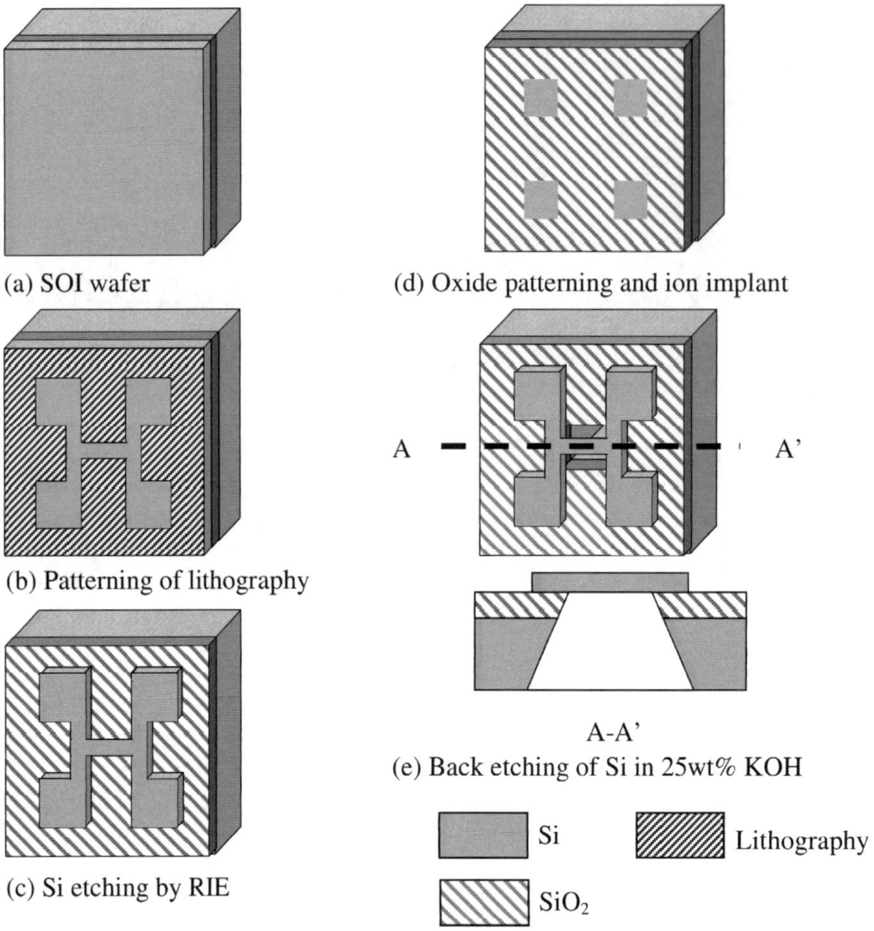

Figure 2. Processes for the fabrication of a free-standing Si thin film.

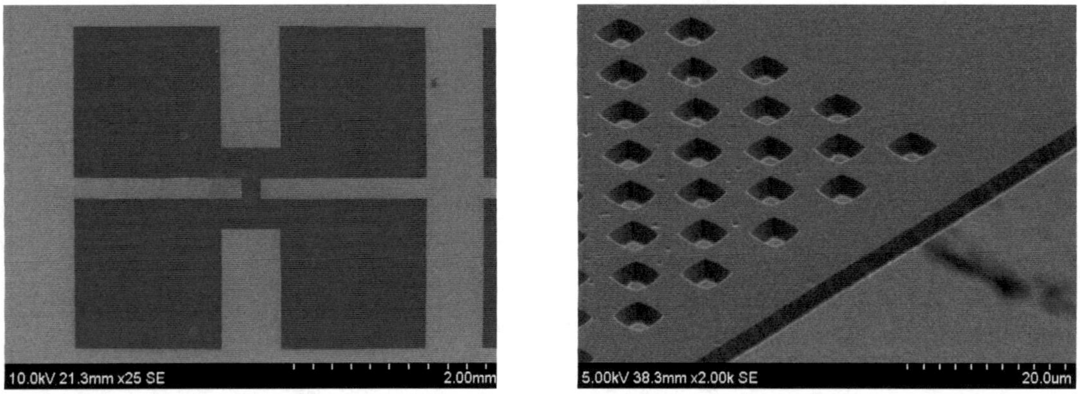

(a) Top view of a free-standing Si thin film. (b) Staggered periodic micropores.

Figure 3. SEM images of a free-standing Si thin film with periodic micropores.

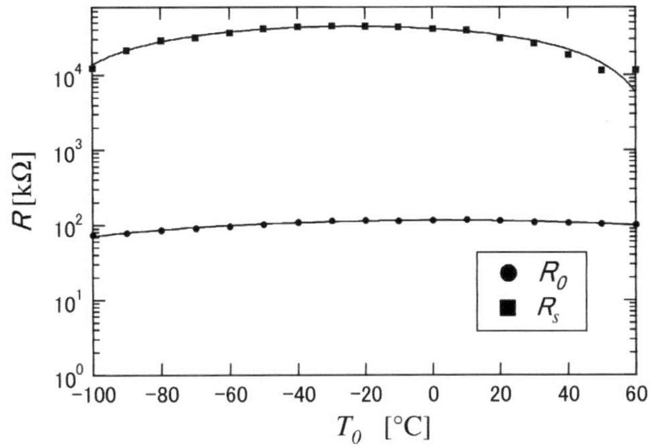

Figure 4. Measured electrical resistance of free-standing Si thin films.

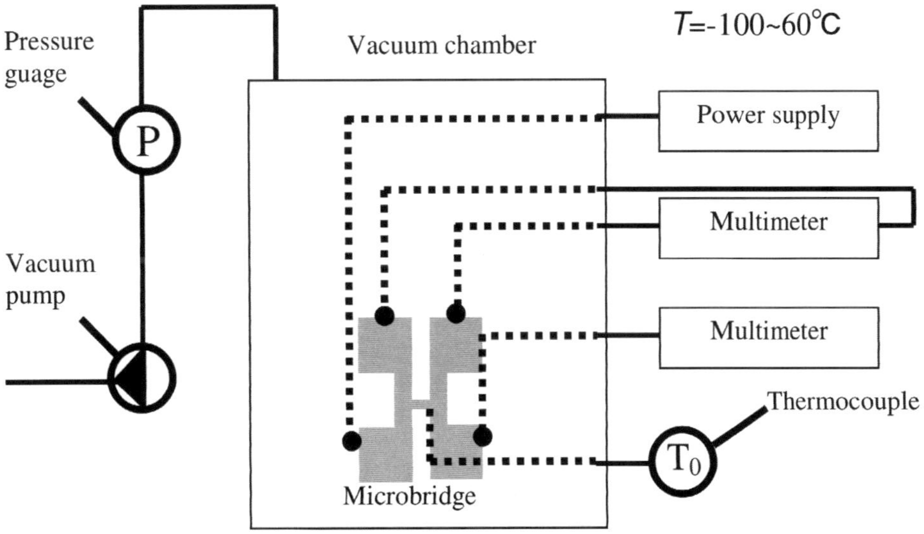

Figure 5. Schematic diagram of the experimental apparatus for thermal conductivity measurement.

4. Results and discussion

Figure 6 shows the in-plane thermal conductivity ratio between the Si thin film with periodic staggered pores and the Si thin film. The ratio is much smaller than that expected based on the classical model (0.6-0.7). Thermal conductivity is determined based on the heat capacity, the phonon group velocity, and the phonon MFP. The heat capacity and the group velocity are constant within the present temperature range. The thermal conductivity of the Si thin film with microholes is much lower than that of the bare Si thin film at low temperature due to ballistic phonon transport [10]. The phonon MFP is artificially reduced by the micropores. The size effect of phonons on the thermal conductivity is significant, even in the microsized structure. The phonon mean free path of silicon calculated by a full spectrum phonon gas model ranged from 1nm to 10mm [9]. The measured thermal conductivity

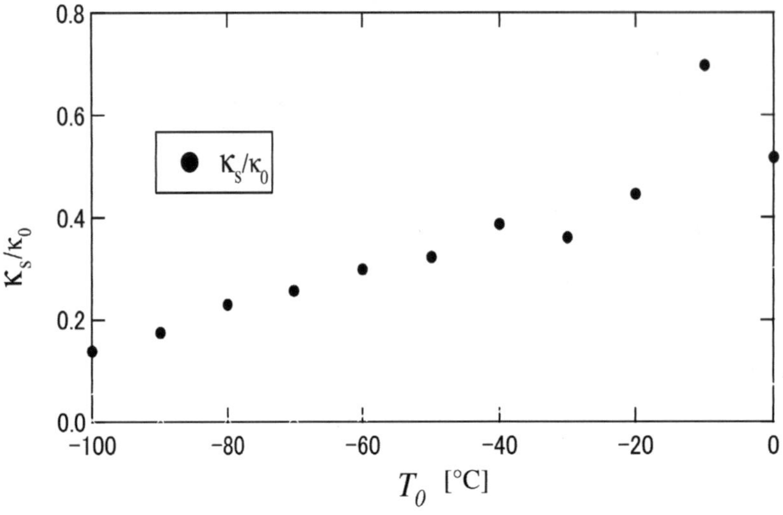

Figure 6. In-plane thermal conductivity ratio between the Si thin film with periodic staggered pores and the Si thin film.

reductions are rational. The size effects on thermal conductivity become more significant at low temperature due to the increasing phonon MFP. The measured thermal conductivity ratio between the Si thin film with periodic straggled pores and the Si thin film was decreased.

The measured electrical conductivity of the Si thin film with micro-holes was too small to discuss the size effects on the transport properties. Mean free path length of electrons is only 110nm in highly doped Si [12,13]. The electron transport might be diffusive, and the effects of microstructures on electrical conductivity were not expected in the present study. Some cracks were observed in the micro-structured samples by SEM. The microcracks were made during the wet chemical etching process due to the residual stress of the thin film. SiN_x thin film was deposited on Si by CVD on the free-standing Si thin films for the protection from the wet chemical etching for the backside. However the SiN_x thin film induces tensile stress. The stress should be compensated by stacking multiple layers to approximate neutral stress. SiO_2 is used for the stress compensation in MEMS process. The sample preparation should be improved to discuss the effects of micro-structures on the electron transport.

5. Conclusion

Free-standing Si thin films with periodic staggered micropores were fabricated and the effective in-plane thermal conductivity was measured. The measured thermal conductivity is much lower than that expected based on the classical model, due to the size effect of phonons, which becomes more significant at lower temperatures. The present results are confirmed by the phonon MFP calculated using a full spectrum phonon gas model. The phonon size effect is non-negligible, even in the microsized structures near room temperature. For the future work, the stress compensation by stacking multiple layers should be introduced to the present fabrication processes to remove the microcracks in the microstructured samples. The effects of different microstructures, such as different hole shape, hole size, and hole arrangements, on transport properties of phonons and electrons will be discussed.

6. Acknowledgment

The authors wish to thank Dr. Hiwada at Fuzzy Logic Systems Institute for his help on microfabrication processes. This work is supported by the Nanotechnology Network of the Ministry of Education, Culture, Sports, Science and Technology (MEXT) of Japan.

7. References

[1] G.E. Moore, 1965, *Electronics Magazine*
[2] C. Dames and G. Chen, *J. Appl. Phys.* **95** 682
[3] D.G. Cahill, K. Goodson, and A. Majumdar, 2002, *ASME J. Heat Trans.* **124** 223
[4] G.J. Snyder and E.S. Toberer, 2008, *Nat. Mater.* **7** 105
[5] L.D. Hicks, and M.S. Dresselhaus, 1993, *Phys. Rev. B* **47** 12727
[6] D.G. Cahill, W.K. Ford, K.E. Goodson, G.D. Mahan, A. Majumdar, H.J. Maris, R. Merlin, and S.R. Phillpot, 2003, *J. Appl. Phys.* **93** 793
[7] G. Chen, and A. Shakouli, 2002, *ASME J. Heat Trans.* **124** 242
[8] A.I. Hochbaum, R. Chen, R.D. Delgado, W. Liang, E.C. Garnett, M. Najarian, A. Majumdar, and P. Yang, 2007, *Nature* **451** 163
[9] C. Dames and G. Chen, 2005, *Thermoelectrics Handbook Macro to Nano*, 42, CRC Press, ed. D. Rowe
[10] X. Zhang, S. Fujiwara, and M. Fuji, 1999, *Int. J. Thermophys.* **21** 4
[11] K. Miyazaki, T. Arashi, D. Makino, and H. Tsukamoto, 2006, *IEEE Trans. Compon. Packag. Technol.* **29** 247
[12] N. W. Ashcroft, and N. D. Mermin, 1976, *Solid State Physics* Chaps 1,2 and 13
[13] S. M. Sze, 1981, *Physics of Semiconductor Devices*. Chap.1
[14] D. Song and G. Chen, 2003, *Appl. Phys. Lett.*. **84** 5

Point Probes: a new generation of magnetic sensors for the measurement of local magnetic fields

M Sheikh Amiri[1], K Szielasko[2], I Altpeter[2], C Boller[1,2], D Fischer[3], J Hahn[3]

[1] Saarland University, Chair of Nondestructive Testing and Quality Assurance, Saarbrücken, Germany
[2] Fraunhofer Institute for Non-Destructive Testing (IZFP), Saarbrücken, Germany
[3]SITEC Industrietechnologie GmbH, Chemnitz, Germany

E-mail: Meisam.SheikhAmiri@izfp.fraunhofer.de

Abstract. In several magnetic Non-Destructive Testing (NDT) methods, the local measurement of the magnetic field inside the material is required. Moreover, looking at difficult part geometries, magnetic field sensors have to be small enough in order to reach the measuring position. The most-used magnetic field sensors are coils, Hall-effect sensors, flux gates and magnetoresistive sensors. However, regarding the industrial application, those sensors are often packaged and cannot be placed close enough to the measuring position.
As part of an ongoing research project funded by the German Ministry of Economics and Technology (BMWi), a new kind of magnetic field sensor was developed and used in order to measure the strength of remanent magnetic field spots. This so-called 'Point Probe' is based upon a needle-shaped ferromagnetic core having a primary coil as a magnetic field source and a secondary coil as an inductive pick-up. This contribution describes the details of the sensor design and its operating principle. The sensitivity of the measured signals for local magnetic fields is described. Finally, a method for nondestructive hardness estimation of materials by using the Point Probe is presented. The results show a high correlation between hardness and a new coercivity-dependent testing parameter.

1. Introduction

Reducing the influence of sample and probe geometry on the measured quantities represents a major challenge in micromagnetic Non-Destructive Testing (NDT). This issue can mostly be attributed to the difficulty of focusing the magnetic field to a certain location. Although it is possible to use miniature yokes in order to overcome this circumstance, their use is restricted to the laboratory environment, where the effects of wear, dust and handling can be controlled. There exists a demand for robust, small probes which obtain highly localized materials information.
In this paper, a novel kind of micromagnetic probe is proposed which is distinguished by a surface-normal magnetic field introduction, only a single pole of the probe being in contact to the material surface. The well-known rules for the transition of electromagnetic field and flux at media boundaries

IOP Conf. Series: Materials Science and Engineering **31** (2012) 012021 doi:10.1088/1757-899X/31/1/012021

demand that the normal component of the magnetic flux density be continuous at the media boundary. If we apply this to our probe design, this means that the flux density in the material immediately below the needle tip is the same as the flux density in the needle tip itself. Since the flux is concentrated in the needle tip, this flux density can reach values close to the saturation of the inspected material's hysteresis curve. The flux density, however, rapidly drops with increasing distance to the contact position. The magnetic circuit is closed through the air, which distinguishes this probe from most conventional probes.
This type of electromagnet introduces an isotropic field to the material, a significant flux density being present in a small volume close to the tip only. Therefore, the influence of the probe orientation and the further surroundings of the measuring position on the magnetization of the investigated spot can be considered negligible.

2. Materials and Experimental Procedure

2.1. Point Probe design
As it can be seen in figure 1, the Point Probe has two coils wound on a ferromagnetic core with a needle-shaped tip. The primary coil (the large one) is connected to a power amplifier which generates the magnetic field. The secondary coil (the small one) is used to pick up the magnetic flux change close to the tip. Since the material of the core has minor effect on the results and to increase the lifetime, a hardened scriber was used.

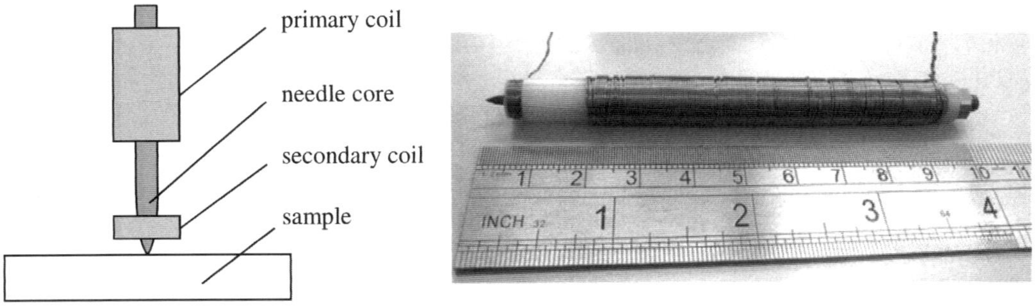

Figure 1: Schematic (left) and real (right) picture of a Point Probe

2.2. Operating modes
Firstly, the situation where the needle tip is in the air is considered. If the primary coil is driven with an alternating voltage of sinusoidal waveform, the voltage induced into the secondary coil reflects the effects of flux change and saturation in the needle, as expected (figure 2, blue curve). However, when ferromagnetic material is contacted with the tip, a sharp voltage peak is observed in each extremum of the signal (figure 2, red curve). The occurrence of this peak can be explained by the flux concentration in the tip, and it reflects very quick transitions from positive saturation to negative saturation of the tip. With the needle tip in the air, the effect of flux-concentration is reduced, and therefore the peak is not visible.

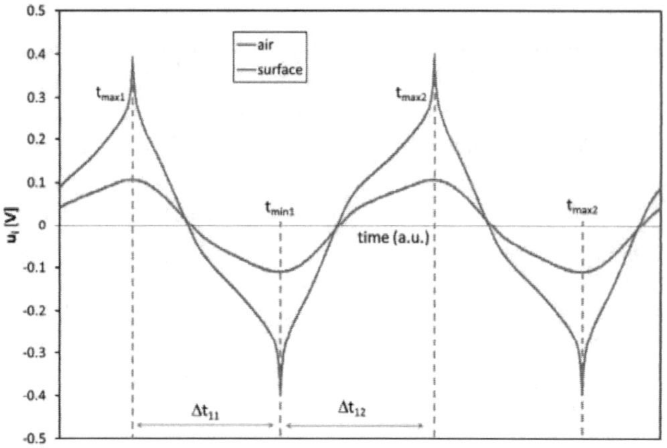

Figure 2: Characteristic waveform of the secondary voltage u_i with the needle tip in the air (blue) and at the surface (red); magnetization frequency: 25 Hz

The following modes of operation can be derived from the specific properties of this signal:

2.2.1. Permeameter mode. It was observed that the height of the sharp peak primarily reflects the local permeability of the material immediately below the tip. Therefore it strongly depends on the magnetic coupling (air gap, coating, etc.) and may be used for the thickness measurement of non-ferromagnetic or magnetically hard coatings.

A quantity V_{peak} was defined which represents the amplitude of the peak in the measured signal. Since the signal can be seen as the superposition of a fundamental, low-frequency signal and the sharp peak signal, algorithms for the decomposition of the signal were sought for. Accordingly, several possible definitions for the determination of V_{peak} are proposed:

- *Vpeak = absolute peak amplitude of the measured, unfiltered signal*: This way, the measured peak value is tainted with the amplitude of the fundamental signal, which is influenced by other ferromagnetic objects close to the probe.
- *Vpeak = peak amplitude after high-pass filtering*: This removes most of the unwanted low-frequency part, but also imposes the filter response to the signal, which can lead to undesirable nonlinearities of the response. In addition, noise can have increasing influence on the result.
- *Vpeak = amplitude of certain higher harmonics*: The occurrence of the peak introduces a series of higher harmonics to the frequency spectrum of the signal. Some of these higher harmonic amplitudes can be used to represent the amplitude of the sharp peak. However, the relation between the spectrum and the material properties is not yet completely understood.
- *Vpeak = quotient between absolute peak and another value of the signal at a fixed time*: This compensates for the amplitude of the low-frequency component, but the result also depends on the signal shape.

Since all definitions proposed above have individual advantages and drawbacks, this paper shows results obtained using the most comprehensible and most simple definition which is mentioned first in the list above and computed as follows:

$$V_{peak} = \frac{\max(u_i) - \min(u_i)}{2} \qquad (1)$$

2.2.2. Magnetometer mode. Theory and Finite Element Method (FEM) simulation indicate that a very small fraction of the material volume below the tip is magnetized close to saturation [1]. This means that the needle partially erases the previous magnetization state in this spot. However, the

IOP Conf. Series: Materials Science and Engineering **31** (2012) 012021 doi:10.1088/1757-899X/31/1/012021

surroundings of the spot still influence the magnetization process and impose an asymmetry to the measured signal. An algorithm for the numerical evaluation of the asymmetry was developed, which allows to determine a measure quantity for the magnetization of the material.

In figure 2, the times corresponding to the minimum and maximum of the sharp peaks in cycle 1 and cycle 2 are denoted t_{min1}, t_{max1}, t_{min2} and t_{max2}, respectively. Based on these quantities, a quantity Δt, which describes the asymmetry of the peak appearance, was calculated as follow:

$$\Delta t = \Delta t_{11} - \Delta t_{12}\ , \quad \text{where } \Delta t_{11} = t_{max1} - t_{min1} \quad \text{and} \quad \Delta t_{12} = t_{min1} - t_{max2} \qquad (2)$$

A software for the automatic data acquisition and determination of Δt was created using National Instruments LabVIEW™. A peak-detector algorithm with spline interpolation of the values close to the peak was used in order to improve the detection accuracy.

2.3. Materials

Eight hardness test reference samples are used in this paper which are shown in figure 3.

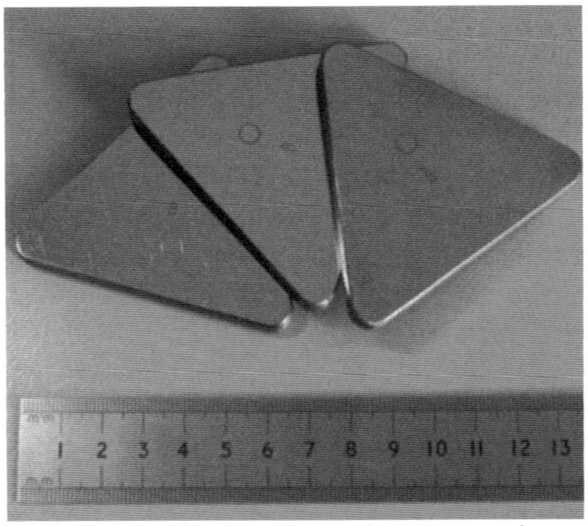

Figure 3: Hardness test reference sample

Also two partially hardened samples were investigated in order to show the suitability of the Point Probe for measuring the local coercivity and consequently the hardness difference. Figure 4 shows the triangle sample and the profile sample which were laser-hardened.

Figure 4: Laser-hardened samples; left: triangle sample which is partially hardened (dark line). Right: profile sample which is hardened on the sides (darker area). The yellow points and lines show the measurement areas.

IOP Conf. Series: Materials Science and Engineering **31** (2012) 012021 doi:10.1088/1757-899X/31/1/012021

3. Results and Discussion

3.1. Hardness estimation

As a part of the BMWi-funded research within the frame of a ZIM research project (see acknowledgements), a method for the verification of proper laser-hardening with low calibration effort has been developed. Unlike most micro-magnetic applications, the main interest is not the quantitative determination of hardness, but rather just the safe detection of poorly hardened spots – independent from the part geometry. The most promising approach found so far is based on the measurement of remanent magnetism after spot-wise magnetization. The approach resembles Förster's "Restpunktpolverfahren" (remanent point pole method) [2] at the first glance, but is distinguished by the completely different way of magnetic field measurement and different kind of application.

In the first step, a magnetic spot was introduced to the material by contacting the surface with a cylindrical Neodymium-Iron-Boron permanent magnet of 3 mm in diameter and height. In the second step, the needle-shaped probe was placed in the center of the magnetized spot in order to measure the normal component of the remanent magnetization of the sample. A precision stand was used in order to guide the magnet and the needle probe. The size of the magnet limits the magnetized area, and the size of the needle tip finally focuses the measurement on a very small spot.

The hardness is clearly reflected in the time difference Δt which was computed according to equation (1). As an example, figure 5 shows the different u_i waveforms for hardness reference discs with 159 HV and 848 HV, in both demagnetized and spot-magnetized state. In the case of the hard-magnetic 848 HV material, the difference between both magnetization states is larger than in the case of the soft-magnetic 159 HV material. The explanation is that hardened material has a higher remanence and therefore retains a higher magnetization than the soft material. The higher magnetization leads to a larger difference between the magnetized and demagnetized state [1, 3], which reflects in a higher Δt.

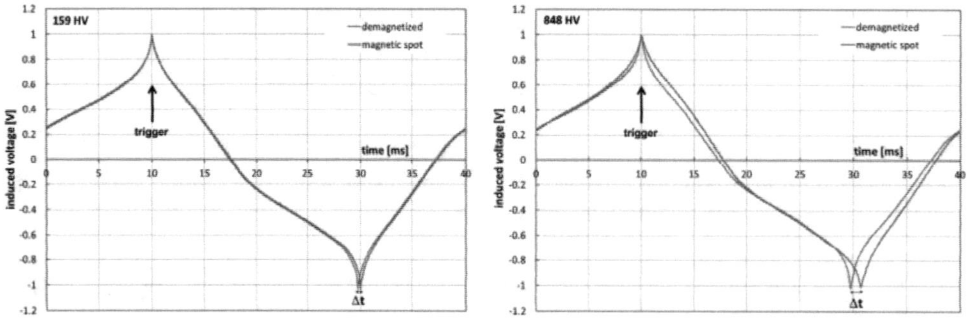

Figure 5: Difference between the u_i waveforms in demagnetized spots (blue curves) and magnetized spots (red curve) of samples with different hardness (left: 159 HV, right: 848 HV, waveforms triggered at t = 10 ms). Magnetization frequency: 25 Hz

In order to demonstrate the hardness estimation of materials with a Point Probe, eight hardness test reference samples were used, because they have homogenous hardness in the whole sample area.

The primary amplitude is one of the most important affecting parameters on the measuring effect size. As it can be seen in figure 6, Δt increases with decreasing primary voltage amplitude. However, the detection of the peaks becomes unstable if the voltage amplitude is too low. As a tradeoff, amplitude of 1 V was chosen for all measurements.

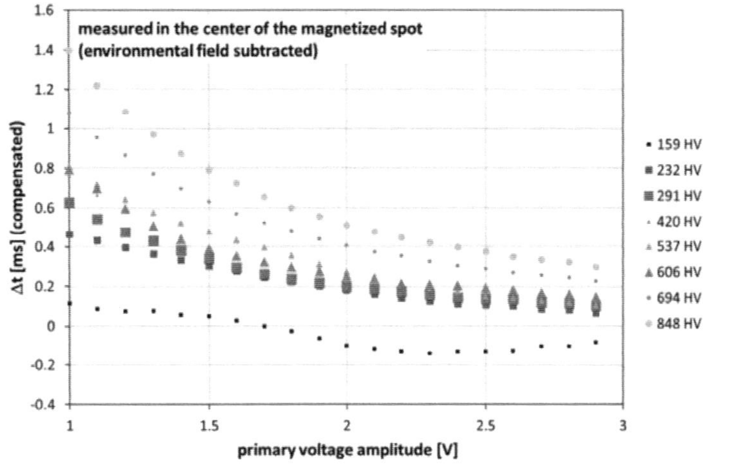

Figure 6: Relation between Δt and primary voltage amplitude on different samples

The hardness dependence of the quantities V_{peak} and Δt was explained more closely at a primary voltage amplitude of 1V.
Figure 7 shows the relationship between V_{peak} and hardness.

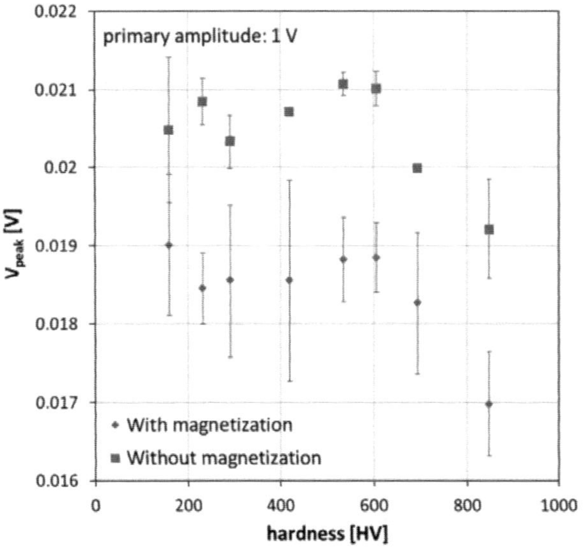

Figure 7: Relation between V_{peak} and hardness with (blue) and without (red) magnetization

As it can be seen, there is no proper relationship between V_{peak} and hardness, neither with magnetization nor without magnetization. As it was found that V_{peak} reflects the permeability of materials and is furthermore strongly affected by surface quality, no clear-cut V_{peak} - hardness correlation could thus be expected. Also, the error bands in samples with magnetic spot are larger in comparison to samples without magnetic spot, because the position of the Point Probe on the magnetic spot affects the V_{peak} value. In comparison to V_{peak} , Figure 8 shows the relationship between Δt and hardness.

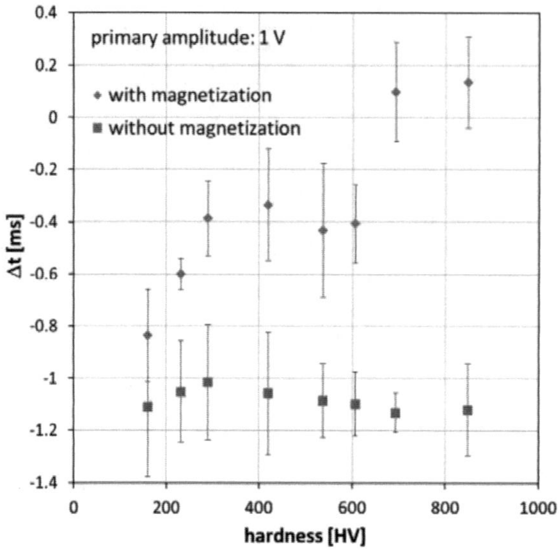

Figure 8: Relation between Δt and hardness with (blue) and without (red) magnetization

Figure 8 indicates the correlation between Δt and hardness for the partially magnetized samples. As it can be seen, two samples with hardness 537 HV and 606 HV deviate from the trend. This might be due to different materials or heat treatment, since the reference samples were made for evaluating conventional hardness tests, not micro-magnetic testing devices. For that reason, all samples were tested with IZFP's 3MA-II (**M**icromagnetic **M**ultiparameters **M**icrostructure and Stress **A**nalyzer) device to measure magnetic properties like the coercivity (H_c) [4]. Figure 9 shows that a linear correlation exists between H_{CM} and Δt, which suggests that the local coercivity can be measured using the Point Probe. Moreover, the two deviating samples were observed to differ in the H_{CM} values as well. This is confirming different magnetic properties.

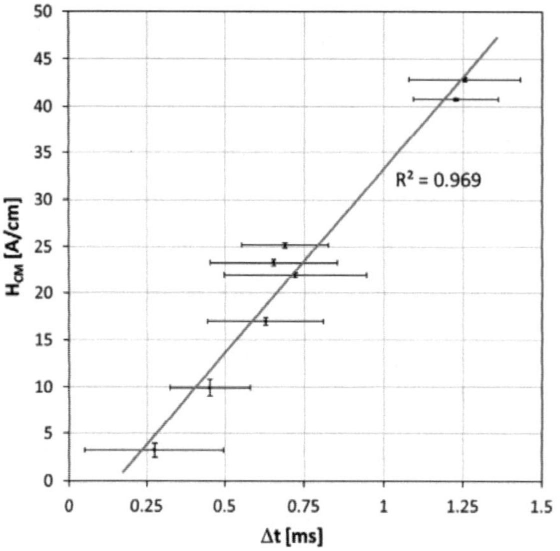

Figure 9: Relation between Δt and H_{CM} which are measured by point probe and 3MA-II, respectively

IOP Conf. Series: Materials Science and Engineering **31** (2012) 012021 doi:10.1088/1757-899X/31/1/012021

3.2. Laser-hardened samples

As the laser hardened depth and area are small, and according to the above mentioned points, it's proved that point probe is measuring the local coercivity field and it's also shown the ability of point probe to estimate hardness of materials. Therefore, the Point Probe would be a proper choice for the investigation of laser hardened samples. Figure 10 and 11 show the Δt values on marked points on the laser hardened parts and non-hardened ones in triangle and profile sample, respectively.

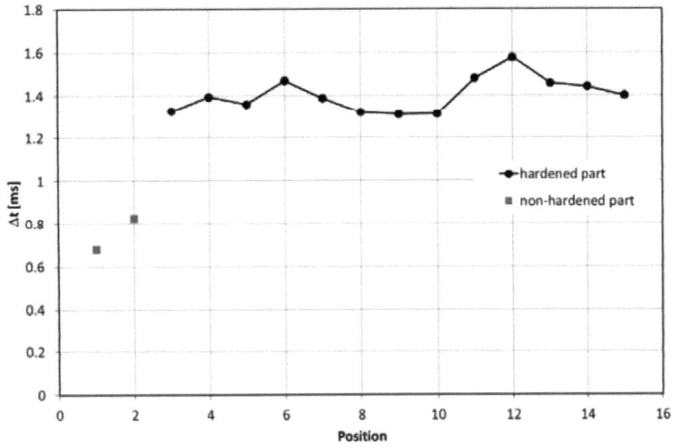

Figure 10: Δt values on hardened (black) and non-hardened (red) sections of the triangle-shaped laser-hardened sample.

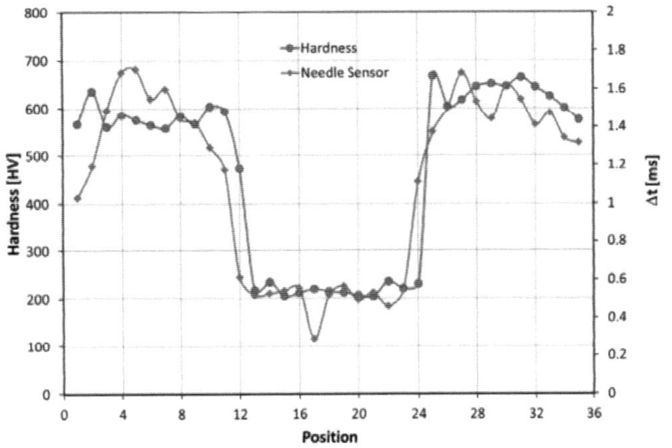

Figure 11: Δt values (blue curve) and hardness profile on hardened (higher values) and non-hardened (lower values in the center) areas of the laser-hardened profile sample

Figure 11 confirms a significant difference between non-hardened and hardened areas and strong correlation between the hardness and Δt. This shows that the point probe could easily recognize the laser-hardened parts from non-hardened ones.

4. Conclusion

In this paper a new micro-magnetic sensor called Point Probe was introduced. Its construction is described and the parameters Δt and V_{peak} are defined. The determination of local magnetic properties was demonstrated. More specifically it was shown that the Point Probe can be used to measure mechanical properties of materials, such as the hardness. The application on two partially laser-hardened samples was successfully demonstrated.

5. Acknowledgment

We would like to thank the German Federal Ministry of Economics and Technology (BMWi) for funding this research within the ZIM programme, project no. KF2455301US9.

6. References

[1] Szielasko K, Sheikh Amiri M, Kopp M, Rabung M and Altpeter I, 'Localized micromagnetic materials characterization with point probes', proceedings of the 9th ICBM, 2011 (to be published).

[2] Hanke E, Prüfung metallischer Werkstoffe, Vol. 2, VEB Deutscher Verlag für Grundstoffindustrie, Leipzig, 1960.

[3] Szielasko K, Sheikh Amiri M, Altpeter I, Boller C, Fischer D and Hahn J, 'Mikromagnetische Härtebeurteilung mit Punktsonden für den industriellen Einsatz bei kleinen Stückzahlen', proceedings of the DGZfP, presented at DGZfP Jahrestagung 2011, Bremen, 2011 (to be published).

[4] Theiner W A, Reimringer B, Kopp H and Gessner M, '3MA-Testing Equipment, Application, Possibilities and Experiences. Characterization of materials', (ed. by P. Höller, V. Hauk, G. Dobmann, C. Ruud, R. Green), Springer-Verlag Berlin, 1989.

IOP Conf. Series: Materials Science and Engineering **31** (2012) 012022 doi:10.1088/1757-899X/31/1/012022

Use of elemental materials for the creation of an in-situ space dust impacts detector

P Faure[1,2], S Matsumoto[1], Y Akahoshi[1], M Cho[1], T Narumi[1], Y Kitazawa[3], A Sakurai[4] and T Koura[1]
[1]Kyushu Institute of Technology, 1-1 Sensui, Tobata, 804-8550 Fukuoka, Japan
[2]École Européenne d'Ingénieurs en Génie des Matériaux, 54000 Nancy, France
[3]IHI Corporation, Japan
[4]Institute for Q-shu Pioneers of Space, 810-0001 Fukuoka, Japan

E-mail: pauline.faure@eeigm.inpl-nancy.fr

Abstract. This research focuses on space dust ranging from 100µm to 1mm. Space dust is mainly due to secondary space debris, which is called ejecta. The objective was to create an inexpensive space dust impacts detector using elemental materials. The detector is a glass/epoxy laminate printed circuit board with an area of 81cm^2 for a weight of 30g. The detector can estimate the number of impacts and can give an approximation of the space dust size. The detector will be mounted on Horyu II that will operate in polar orbit for one year. In this article the authors report: a) the production of ejecta, b) the ejecta experiments on solar array coupon, aluminium honeycomb and CFRP/aluminium honeycomb, c) the detector's working principle and d) the estimations of the minimum detectable size of debris and collision probability. The ejecta experiments demonstrated that the ejecta's mass is 7 to 46 times higher than the projectile's mass. For space dust in the range 100µm – 600µm in diameter, the collision probability was calculated to be 16.5 percent. The detector's capabilities to detect broken lines and to transmit the data to the on-board computer were also demonstrated. This in-situ space dust impacts detector is thus a very promising research area for its lightness, low cost and its ability to provide immediate data on space dust population.

1. Introduction

1.1. Space debris situation

Space debris is non-functional man-made space objects in Earth's orbit or re-entering the Earth's atmosphere [1-3]. In 2005, Gelhaus [4] estimated that about 5,800,000,000,000 objects characterized as space debris were orbiting the Earth with a size ranging from 100µm to several meters. Space debris is a pending situation needing to be resolved in order to maintain the well being of current and future missions. This article focuses on small-size space debris, called space dust, ranging from 100µm to 1mm, which is orbiting in low Earth orbit (LEO). The Committee on Space Debris *et al.* [5] warned

[1] To whom any correspondence should be addressed.

that despite their small size, the space dust can be harmful to spacecraft and may result in the termination of a mission.

1.2. Space debris detection systems

Space dust is very difficult to observe using ground observations like radars and optical telescopes, as they are limited to 10cm space debris in LEO [1]. The only data available for the quantification of space dust is given by the analysis of retrieved surfaces. This technique has two main drawbacks: a long time is needed to access the data and not enough surfaces were retrieved to perform statistical analysis.

To overcome time and size issues, in-situ detection systems were developed. The space dust impacts detector proposed in this article belongs to this category and has the advantage to be low power consuming, 0.01W, light, 30g, low cost, about EUR180 (JPY20,000; USD250) and to provide instant data on micrometer size space debris.

2. Ejecta

Ejecta are secondary debris produced when primary space debris impacts the surface of a spacecraft. Ejecta are the main contributors to the population of space dust in LEO.

2.1. Matter ejection under hypervelocity impact

Matter is ejected at hypervelocity impacts in a variety of ways (Figure 1). Jetting consists in the ejection of small and fast liquid particles at grazing angles. Jetting represents less than 1 percent of the total ejected mass. Cone consists in the ejection of small and fast solid particles at an angle of about 60°. Spall consists in the ejection of large fragments at low velocities perpendicularly to the surface, but occurs only with brittle targets. The ejected matter is collectively called ejecta.

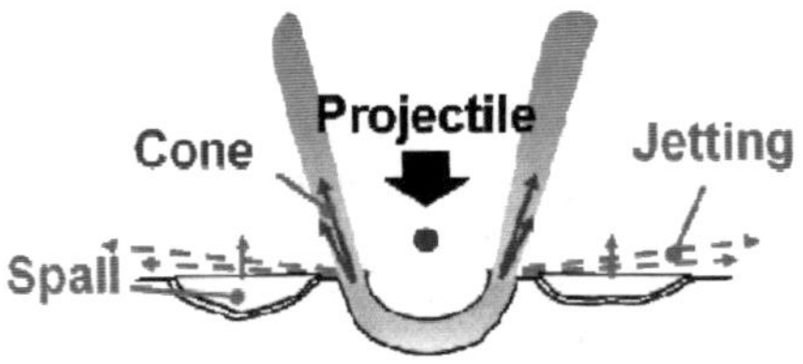

Figure 1. Schematic representation of matter ejection under hypervelocity impact.

2.2. Ejecta production comparison

Experiments on solar array coupon, aluminium honeycomb and carbon fibres reinforced polymer (CFRP)/aluminium honeycomb were performed to observe the differences in ejecta production. Experiments were carried out at Kyushu Institute of Technology (KIT) using a two-stage light gas gun (2SLGG) (Figure 2).

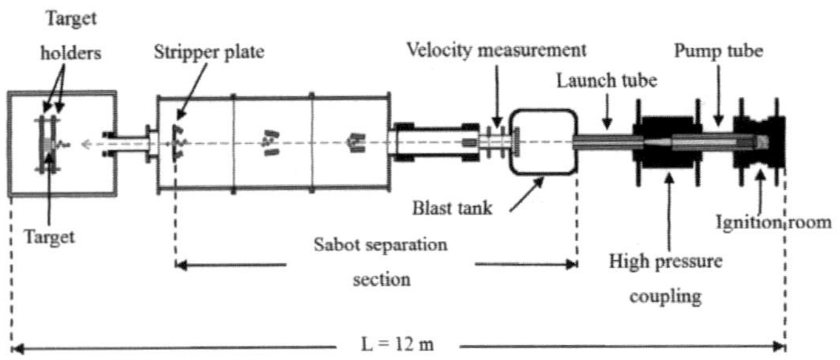

Figure 2. Schematic of KIT's two-stage light gas gun.

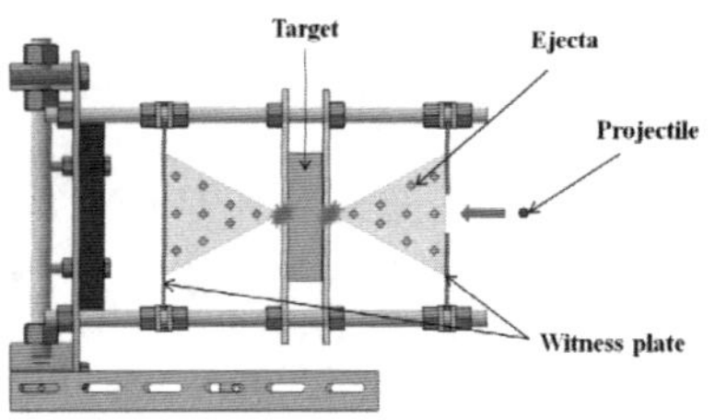

Figure 3. Close-up on target set-up.

For each experiment a 1mm diameter aluminium sphere, Al2017, was launched at about $5km.s^{-1}$. The created ejecta impacted the witness plates that were then further analysed to estimate the number of ejecta (Figure 3). Table 1 demonstrates that depending on the material, the mass of ejecta is 7 to 46 times higher than the projectile mass.

Table 1. Results of the ejecta production experiments.

	Target Material		
	Solar Array Coupon	**Aluminium Honeycomb**	**CFRP/Aluminium Honeycomb**
Projectile Velocity [$km.s^{-1}$]	5.37	4.79	4.97
Projectile Mass [mg]	1.5	1.5	1.6
Ejecta Mass [mg]	68.8	10.3	41.1
Ratio Ejecta Mass/Projectile Mass	**45.9**	**6.9**	**25.7**

Solar array coupons, aluminium honeycomb and CFRP/aluminium honeycomb are widely used materials in the aerospace industry. The experiments demonstrated that they produced a non-negligible amount of ejecta when impacted by space debris. In space, this phenomenon increases small-size space debris population, making the available space in LEO more crowded and the amount of space debris will eventually be detrimental to current and future space missions. Moreover, since solar array coupons, aluminium honeycomb and CFRP/aluminium honeycomb can be considered as advanced materials, they are expensive. To satisfy our purposes, the use of the previously described materials had therefore no beneficial advantage and an elemental glass/epoxy laminate printed circuit board (PCB) was chosen to create the space dust impacts detector.

3. Space dust impacts detector

3.1. Short story

The space dust impacts detector proposed in this article is derived from an original idea as a result of the collaboration between IHI Corporation (IHI) and the Institute of Q-shu Pioneers of Space (iQPS). Kitazawa *et al.* [6, 7] describe the sensor as being active, flexible, low in weight and in power consumption and able to provide real time data. IHI and iQPS's debris sensor is composed of 8 flexible bands, each containing 125 copper lines for a total of 1000 conductive lines forming the detecting pattern. A line is 50µm wide, the distance between them is 100µm and the thickness of each band is 25µm [8]. The sensor's total area is 270mm x 190mm.

The original purpose was to mount the detector described above on the nano-satellite Horyu II built at KIT; however, due to budget and dimension constraints, the detector developed by IHI and iQPS did not fulfil the requirements; thus, it was modified. The modified detector named Q-shu Debris Sensor (QDS) is described below.

3.2. Q-shu Debris Sensor's working principle

3.2.1. Overview. The detector is a square PCB with each side measuring 90mm for a mass of 30g. The detector's front face, layer 4, is exposed to the space environment (Figure 4). On the front face is the detecting pattern formed by the 128 copper lines. Each line is 127µm wide and the distance between two lines is 190µm. The rear face, layer 1, is inside the satellite (Figure 5). The rear face contains the electronic components: four 32 channels multiplexers and a PIC 16F877A microcontroller. PCB's total thickness is 609µm and is composed of four layers.

IOP Conf. Series: Materials Science and Engineering **31** (2012) 012022 doi:10.1088/1757-899X/31/1/012022

Figure 4. QDS flight model front face with the 128 lines forming the detecting pattern.

Figure 5. QDS flight model rear face with the electronic components.

QDS's objective is to monitor the change of voltage induced by space dust impacts and from this, to estimate the number of impacts. More precisely, when space dust will impact the board, one or several lines will be cut. This will result in a voltage change that will be monitored by the electronic components, which will then transmit the data to the on-board computer (OBC). Depending on the number of broken lines, the detector is also able to estimate the size of the impacting space dust.

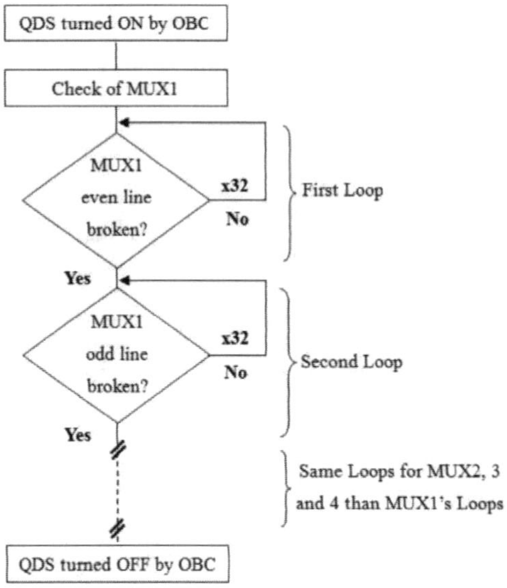

Figure 6. Flowchart describing the programme implemented in the PIC microcontroller.

3.2.2. Programming. To analyze the state of each line, the PIC microcontroller was programmed. The programme states that for each multiplexer, even and odd lines are verified (Figure 6). An even line is defined as being a line, which PIC microcontroller's address ends with an even number. On the other hand, an odd line is defined as being a line, which PIC microcontroller's address ends with an odd number. When a broken line is found, the loop is disrupted, and the programme goes to the next loop.

Data is only sent to the OBC when a broken line is encountered. By default, the programme supposes that all lines are initially not broken. The total data size is 1 byte.

4. Experimental setup

QDS had to undergo a series of tests in order to be accepted on board of Horyu II. The tests can be divided into two categories: hypervelocity impact (HVI) test and environmental tests. Environmental testing includes thermal shock, thermal vacuum and vibrations tests.

HVI, thermal shock and vibrations tests experimental setups were previously described [9] for the QDS engineering model (EM). These tests' setups are therefore not reported in this paper. Concerning the thermal vacuum test, it was performed on the QDS pre-flight model (pre-FM) in a LEO chamber with the vacuum fixed at $1.33x10^{-4}$Pa, the low temperature fixed at -20°C and the high temperature fixed at +55°C.

5. Results and discussion

The results regarding the QDS EM HVI, thermal shock and vibrations tests were already reported [9] and will not be discussed in this paper.

The newest version of the Meteoroid and Space Debris Terrestrial Environment Reference (MASTER) model, named MASTER-2009, was released during the QDS flight model (FM) development changing the previous results [9] obtained for the collision probability. In this section, the results from the thermal vacuum test will be discussed and the collision probabilities calculated by the MASTER-2005 model and by the MASTER-2009 model will be compared.

5.1. Hypervelocity impact tests

5.1.1. Ejecta. For the experiment with Al_2O_3, there were about 300 particles launched at $4.8km.s^{-1}$, whereas for the SUS304 experiment, there were about 250 particles launched at $4.1km.s^{-1}$. For both experiments, the pressure was equal to 6kPa in both sabot separation section and target chamber. Table 2 shows the results of the ejecta production with Al_2O_3 projectiles. Even though projectile differs from the one used with solar array coupon, aluminium honeycomb and CFRP/aluminium honeycomb materials, it should be noted that the ejecta mass relatively to the projectile mass is much lower.

Table 2. Results of the ejecta production experiment with the printed circuit board.

	Printed Circuit Board as Target material
	Al_2O_3 Projectile
Projectile Velocity [$km.s^{-1}$]	4.8
Projectile Mass [mg]	0.1439
Ejecta Mass [mg]	0.4309
Ratio Ejecta Mass/Projectile Mass	**2.99**

5.1.2. Collision probability. First, the impact flux, q, has to be calculated with the MASTER model. From there, n can be estimated using equation (1). Finally, the collision probability can be estimated using equation (2).

$$n = qAt \tag{1}$$

A, area exposed to the space environment [m^2]; *t*, time of exposure to the space environment [year].

$$\text{Collision probability} = 1 - \text{PNP} = 1 - e^{-n} \tag{2}$$

PNP, probability of no penetration.

The impact flux was calculated with both MASTER-2005 and 2009 models for particles ranging from 100μm to 1mm. The number of impacts was calculated for a surface area of 22cm^2 corresponding to the surface of the detecting pattern only. With the MASTER-2005 model, the simulation could not have been performed for years after 2005; thus, to compare both models, the period 2000-2001 was arbitrarily chosen. Calculations with MASTER-2009 were also performed for the time period 2012-2013, which corresponds to the time during which Horyu II will orbit. Results are shown in Table 3.

Table 3. The impact flux, q, and the number of impacts, n, as calculated with the MASTER-2005 and 2009 models for particles ranging from 100μm to 1mm.

MASTER-2005	Period	2000-2001	q	**83.28**
			n	**0.1832**
MASTER-2009	Period	2000-2001	q	**77.84**
			n	**0.1712**
		2012-2013	q	**82.27**
			n	**0.1810**

Results show that there is a difference of about 6.5 percent between the two models. This could be explained by the fact that in the MASTER-2009 model, the number of NaK droplets was decreased from 8kg to 5.3kg because the model considers that only 30,000 droplets with a diameter superior to 5mm are still on orbit [10]. When the MASTER-2009 model results are compared to each other it appears that there is an increase of about 5.7 percent between 2000-2001 and 2012-2013. It illustrates the fact that the 2009 version includes the last space debris created from the collision of Iridium with Cosmos and from the fragmentation of Feng-Yun in 2009 and 2007, respectively.

Previous HVI testing [9] demonstrated that the sensor will be perforated by steel particles as small as 355μm, which energetically corresponds to alumina particles with a diameter of 576μm. To obtain the collision probability, q and n have been recalculated for particles ranging from 100μm to 600μm. Results are gathered in Table 4.

Table 4. The impact flux, q, and the number of impacts, n, as calculated with the MASTER-2005 and 2009 models for particles ranging from 100μm to 600μm.

MASTER-2005	Period	2000-2001	q	**83.19**
			n	**0.1830**
MASTER-2009	Period	2000-2001	q	**77.78**
			n	**0.1711**
		2012-2013	q	**82.20**
			n	**0.1808**

The collision probability, equation (3), was estimated using q and n calculated for the 2012-2013 period.

$$\text{Collision probability} = 1 - e^{-0.1808} = 0.1654 \quad (3)$$

In respect of its detecting pattern area and mission time, the sensor has more than 16 percent chance to be hit by space dust in the range 100μm - 600μm in diameter. This is thus very promising to acquire in-situ data on the space dust population.

5.2. Thermal vacuum test

Test verifications consisted in two main points. First point was to check visually that QDS was not affected by the test by making sure no components detached or partially melted. Second point was to check that the output data displayed on a computer screen are the same before and after the testing. To achieve it, lines had to be cut prior to the test. For obvious reason this was not performed on the FM but, on the pre-FM. The results were in agreement with each other before and after the test. Figure 7 shows captures of the results as displayed on a computer screen.

a)
b)

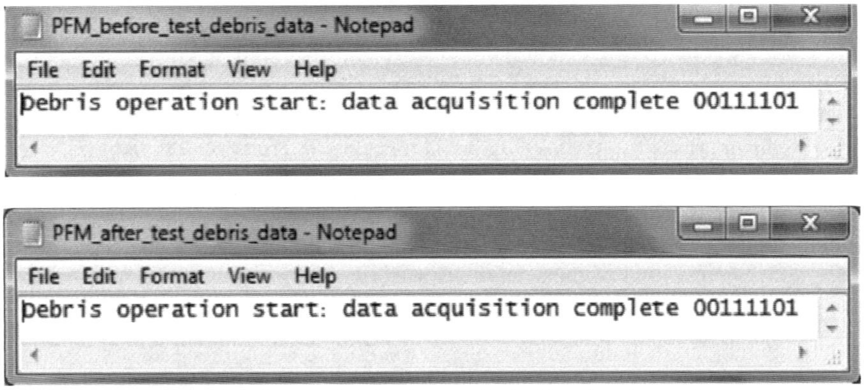

Figure 7. Pre-FM results obtained a) before and b) after the thermal vacuum test.

6. Conclusions

IHI Corporation and Institute for Q-shu Pioneers of Space's debris impacts sensor was modified to be adapted on the nano-satellite Horyu II under development at Kyushu Institute of Technology. The modified sensor was named Q-shu Debris Sensor. The sensor is a square glass/epoxy laminate printed circuit board with each side measuring 90mm for a total weight of 30g. On the front face is the pattern formed by the copper lines, which constitute the detection area. On the rear face are the electronic components, which control each line state and then send the data to the on-board computer, which will finally transmit the data to a ground station.

The different environment tests proved the capability of the detector to withstand the harsh spatial environment, whereas the hypervelocity impact test demonstrated the detector's ability to have its detecting lines broken by micrometer size debris. Besides, the collision probability was calculated using the latest version of the MASTER model showing that there are more than 16 percent chances the detector will be impacted by space dust with a size ranging from 100μm to 600μm.

The in-situ space dust impacts detector made from a glass/epoxy laminate printed circuit board is thus a very promising system, which is low in weight and cost and which is capable of providing instant data on the space dust population.

7. Acknowledgements

The authors would like to sincerely thank Haruhisa Matsumoto from the Japanese Aerospace Exploration Agency for his participation as well as his help during the development of the detector. The authors are also grateful to Hiroki Nishimura and all Horyu members for their great support and continuous help during this project.

8. References

[1] Klinkrad H 2006 *Space Debris: Models and Risk Analysis* (Berlin: Springer)
[2] Bariteau M 2001 *Proliferation des Debris Orbitaux: Production et Evolution des Particules Secondaires* Departement d'Environnement Spatial de l'ONERA (Toulouse)
[3] Working Group 4 2007 *IADC Space Debris: Mitigation Guidelines* (IADC)
[4] Gelhaus J 2010 *61st Int. Astronautical Congress (Prague, Czech Republic, 27 September to 1 October 2010)* Vol. 4 p 2996
[5] Committee on Space Debris, Aeronautics and Space Engineering Board, Commission on Engineering and Technical Systems and National Research Council 1995 *Orbital Debris: a Technical Assessment* (Washington: National Academy Press)

[6] Kitazawa Y, Matsumoto H, Sakurai A, Yasaka T, Funakoshi K and Hanada T 2009 Development of a new type sensor for micrometeoroid and space debris in-situ measurement at JAXA 2009-r-2-37p

[7] Kitazawa Y *et al* 2008 Current problems of meteoroid and debris environment models and related studies of JAXA EGU2008-A-01753

[8] Kitazawa Y, Matsumoto H, Sakurai A, Yasaka T, Funakoshi K, Hanada T, Hasegawa S and Kadono T 2010 Status report on the development of a sensor for in-situ space dust measurement *38th COSPAR Scientific Assembly* (Bremen)

[9] Faure P, Matsumoto S, Akahoshi Y, Cho M, Narumi T, Kitazawa Y, Sakurai A and Koura T 2011 *Proc. Int. Symposium on Space and Technology Science* (Okinawa) Japan Society for Aeronautical and Space Sciences

[10] Wiedemann C, Flegel S, Gelhaus J, Moeckel M, Klinkrad H, Krag H and Voersmann P 2011 The space debris environment model MASTER-2009 28th ISTS (Okinawa) 2011-r-22